T0155866

Optical Communications

Martin Sibley

Optical Communications

Components and Systems

Third Edition

Martin Sibley
Cardiff, UK

ISBN 978-3-030-34358-3   ISBN 978-3-030-34359-0   (eBook)
https://doi.org/10.1007/978-3-030-34359-0

This Springer imprint is published by the registered company Springer Nature Switzerland AG
The registered company address is: Gewerbestrasse 11, 6330 Cham, Switzerland

Dedicated to Magda and Emily

Preface to the Third Edition

It is some years since the second edition of this book was published and great strides have been made in the field of high-speed optical fibre communications in the interval. We are now seeing optical fibre links with an aggregate data rate in excess of 1 Tbit/s, single-mode lasers with linewidths measured in MHz and less, surface-emitting lasers that can be tested before being packaged, electronics capable of working with 40 Gbit/s data, heterodyning techniques with tunable lasers, in-line optical amplifiers, graphene detectors with very wide spectra range and very high speed and free-space optical communications. As these advances become common-place, so they have been incorporated in the main text rather than being put into a future development section. You will find that the introductory, fundamental parts are the same as for the second edition. This is because a knowledge of the funda-mentals is essential to the understanding of advanced techniques.

The speed at which advances are being made is breathtaking. This is being driven by a need to increase capacity due to a combination of factors – video on demand, search engines, mobile telephones and the Internet of Things. All of these factors, plus others, mean that aggregate data rates are set to rise beyond a Tbit/s. One thing that is certain, optical communications will be at the forefront of the information age. Enjoy reading the book.

My thanks go to my family for their encouragement and support.

Anyone who stops learning is old, whether at twenty or eighty. Anyone who keeps learning stays young.
Henry Ford

Cardiff, UK   Martin Sibley

Contents

1 Introduction 1
   1.1 Historical Background 1
   1.2 The Optical Communications Link 4
   Recommended Reading 7

2 Optical Fibre 9
   2.1 Propagation of Light in a Dielectric 10
       2.1.1 The Wave Equation 11
       2.1.2 Propagation Parameters 13
       2.1.3 Group Velocity and Material Dispersion 17
   2.2 Propagation in a Planar Dielectric Waveguide 22
       2.2.1 Reflection and Refraction at Boundaries 22
       2.2.2 Propagation Modes: Ray-Path Analysis 29
       2.2.3 Propagation Modes: Modal Analysis 32
       2.2.4 Modal Dispersion: Ray-Path Analysis 40
       2.2.5 Modal Dispersion: Modal Analysis 42
       2.2.6 Waveguide Dispersion: Ray-Path and Modal Analysis 43
       2.2.7 Numerical Aperture 45
   2.3 Propagation in Optical Fibres 46
       2.3.1 Propagation in Step-Index Optical Fibres 47
       2.3.2 Dispersion in Cylindrical Waveguides 54
       2.3.3 Step-Index Multimode Fibre 57
       2.3.4 Step-Index Single-Mode Fibre 61
       2.3.5 Graded-Index Fibre 64
   2.4 Calculation of Fibre Bandwidth 65
   2.5 Attenuation in Optical Fibres 68
       2.5.1 Impurity Absorption 68
       2.5.2 Rayleigh Scattering 69
       2.5.3 Material Absorption 70
       2.5.4 Electron Absorption 70

2.5.5   PCS and All-Plastic Fibres   70
2.6   Fibre Materials and Fabrication Methods   71
2.6.1   Materials   71
2.6.2   Modified Chemical Vapour Deposition (MCVD)   71
2.6.3   Fibre Drawing from a Perform   72
2.6.4   Fibre Drawing from a Double Crucible   73
2.7   Connectors and Couplers   74
2.7.1   Optical Fibre Connectors   74
2.7.2   Optical Fibre Couplers   75
Recommended Readings   78

3   Optical Transmitters   79
3.1   Semiconductor Diodes   79
3.1.1   Intrinsic Semiconductor Material   80
3.1.2   Extrinsic Semiconductor Material   83
3.1.3   The p-n Junction Diode Under Zero Bias   86
3.1.4   The p-n Junction Diode Under Forward Bias   92
3.2   Light Emission in Semiconductors   98
3.2.1   Direct and Indirect Band-Gap Materials   98
3.2.2   Rate Equations   100
3.3   Heterojunction Semiconductor Light Sources   106
3.4   Light-Emitting Diodes (LEDs)   108
3.4.1   Surface-Emitting LEDs   109
3.4.2   Edge-Emitting LEDs (ELEDs)   110
3.4.3   Spectral Characteristics   110
3.4.4   Modulation Capabilities and Conversion Efficiency   111
3.5   Semiconductor Laser Diodes (SLDs)   113
3.5.1   Stimulated Emission   115
3.5.2   Spectral Characteristics   125
3.5.3   Modulation Capabilities   129
3.5.4   SLD Structures   133
3.6   Solid-State and Gas Lasers   135
3.6.1 Nd^3 +:YAG Lasers   136
3.6.2   HeNe Lasers   138
3.7   Light-Wave Modulation   139
3.7.1   LED Drive Circuits   139
3.7.2   SLD Drive Circuits   140
3.7.3   External Modulators   142
3.8   Fibre Lasers   151
Recommended Reading   152

4   Photodiodes   153
4.1   V-I Characteristics of Photodiodes   153
4.2   Photoconduction in Semiconductors   155
4.2.1   Photon Absorption in Intrinsic Material   155

        4.2.2   Photon Absorption in Reverse-Biased p-n Diodes 157
   4.3   PIN Photodiodes 162
        4.3.1   Structure 163
        4.3.2   Depletion Layer Depth and Punch-Through Voltage 164
        4.3.3   Speed Limitations 165
        4.3.4   Photodiode Circuit Model 167
        4.3.5   Long-Wavelength PIN Photodiodes 168
   4.4   Avalanche Photodiodes (APDs) 169
        4.4.1   APD Structures 170
        4.4.2   Current Multiplication 172
        4.4.3   Speed Limitations 173
   4.5   Metal Semiconductor Metal (MSM) Photodiodes 174
   4.6   Photodiode Noise 174
        4.6.1   PIN Photodiode Noise 175
        4.6.2   APD Noise 177
   Recommended Reading 181

5   Introduction to Receiver Design 183
   5.1   Fundamentals of Noise Performance 183
   5.2   Digital Receiver Noise 186
        5.2.1   Raised-Cosine Spectrum Pulses 187
        5.2.2   Determination of I_2 and I_3 190
        5.2.3   Statistical Decision Theory 191
        5.2.4   Photodiode Noise 197
        5.2.5   Timing Extraction 200
   5.3   Analogue Receiver Noise 202
   5.4   Comparison of APD and PIN Receivers 204
   5.5   Measurement and Prediction of Receiver Sensitivity 205
        5.5.1   Measurement of Receiver Sensitivity 205
        5.5.2   Prediction of Receiver Sensitivity 206
   Recommended Reading 207

6   Preamplifier Design 209
   6.1   High Input Impedance Preamplifiers 210
        6.1.1   Frequency Response 211
        6.1.2   Noise Analysis 213
        6.1.3   Dynamic Range 214
        6.1.4   Design Example 215
   6.2   Transimpedance Preamplifiers 216
        6.2.1   Frequency Response 218
        6.2.2   Noise Analysis 220
        6.2.3   Dynamic Range 221
        6.2.4   Design Example 222
   6.3   Common-Collector Front-End Transimpedance Designs 223
        6.3.1   Frequency Response 224

     6.3.2   Noise Analysis... 225

     6.3.3   Design Example... 225

  6.4 Bootstrapped Common-Collector Front-End Transimpedance

       Designs... 228

  Recommended Reading... 228

7 Current Systems and Future Trends........................... 229

  7.1 System Design... 229

  7.2 Current Systems... 231

  7.3 Long-Haul High-Data-Rate Links........................... 233

     7.3.1   Optical Fibre Transmission Bands..................... 233

     7.3.2   Advanced Modulation Techniques.................... 234

     7.3.3   Fibre Amplifiers.................................. 235

     7.3.4   Coherent Detection............................... 236

     7.3.5   Wideband Preamplifiers........................... 237

     7.3.6   Optical Solitons................................. 238

  7.4 Free-Space Optical Communications....................... 240

  7.5 Future Trends... 240

     7.5.1   Fluoride-Based Optical Fibres..................... 240

     7.5.2   Graphene Detectors.............................. 241

     7.5.3   Optical Wireless................................ 242

     7.5.4   Crystalline Fibres............................... 242

     7.5.5   Spatial Division Multiplexing (SDM)................ 242

     7.5.6   Passive Optical Networks (PONs)................... 243

  Recommended Reading... 244

Index.. 245

List of Symbols

α	Attenuation constant/absorption coefficient
α_e	Electron ionisation coefficient
α_h	Hole ionisation coefficient
$\mathbf{a_x}, \mathbf{a_y}, \mathbf{a_z}$	Unit vectors
A_o	Voltage gain
$A(\omega)$	Complex voltage gain
β	Phase constant
b	Binding parameter of modes in a waveguide
B	Bit rate in digital or bandwidth in analogue systems
B_{eq}	Noise equivalent bandwidth
c	Velocity of light in a vacuum (3×10^8 m/s)
C_π	Base-emitter capacitance
C_d	Total diode capacitance
C_f	Parasitic capacitance of feedback resistor
C_{gd}	Gate-drain capacitance
C_{gs}	Gate-source capacitance
C_{in}	Preamplifier input capacitance
C_j	Junction capacitance
C_s	Stray capacitance
δn	Fractional refractive index difference
δE_c	Conduction band step
δE_v	Valence band step
D_{mat}	Material dispersion coefficient
D_n	Diffusion coefficient for electrons
D_h	Diffusion coefficient for holes
D_{wg}	Waveguide dispersion coefficient
D_{pmd}	Polarisation mode dispersion coefficient
ε_o	permittivity of free space (8.854×10^{-12} F/m)
ε_r	Relative permittivity
E	Electric field strength

E_c	Conduction band energy level
E_f	Fermi level
E_g	Band-gap difference
E_v	Valence band level
$F(M)$	Excess noise factor
γ	Propagation coefficient
g	Gain per unit length
g_m	Transconductance
h	Planck's constant (6.624×10^{-34} J s)
$h_f(t)$	Pre-detection filter impulse response
$h_{out}(t)$	Output pulse shape
$h_p(t)$	Input pulse shape
H	Magnetic field strength
$H_{eq}(\omega)$	Equalising network transfer function
$H_f(\omega)$	Pre-detection filter transfer function
$H_{out}(\omega)$	Fourier transform (FT) of output pulse
$H_p(\omega)$	FT of received pulse
$H_T(\omega)$	Normalised transimpedance
$\langle i_n^2 \rangle_0$	Mean square (m.s.) noise current for logic 0 signal
$\langle i_n^2 \rangle_1$	m.s. noise current for logic 1 signals
$\langle i_n^2 \rangle_c$	m.s. equivalent input noise current of preamplifier
$\langle i_n^2 \rangle_{DB}$	m.s. photodiode bulk leakage noise current
$\langle i_n^2 \rangle_{DS}$	m.s. photodiode surface leakage noise current
$\langle i_n^2 \rangle_{pd}$	m.s. photodiode noise current
$\langle i_n^2 \rangle_Q$	Quantum noise
$\langle i_n^2 \rangle_T$	Total signal-independent m.s. noise current
$\langle i_s^2 \rangle$	m.s. photodiode signal current
$i_s(t)$	Photodiode signal current
I_2, I_3	Bandwidth type integrals
I_b	Base current
I_c	Collector current
I_d	Total dark current
I_{diode}	Total diode current
I_g	Gate leakage current
I_M	Multiplied diode current
I_{max}	Maximum signal diode current
I_{min}	Minimum signal diode current
I_s	Signal-dependent, unmultiplied photodiode current
$\langle I_s \rangle$	Average signal current
$\langle I_s \rangle_0$	Average signal current for logic 0
$\langle I_s \rangle_1$	Average signal current for logic 1
I_{th}	Threshold current
I_{DB}	Photodiode bulk leakage current
I_{DS}	Photodiode surface leakage current

ISI	Inter-symbol interference
J	Current density
J_{th}	Threshold current density
k	Boltzmann's constant (1.38×10^{-23} J/K)
k_o	Free-space propagation constant of a propagating mode
λ	Wavelength
L_n	Diffusion length in n-type material
L_p	Diffusion length n p-type material
m	Modulation depth
M	Multiplication factor
M_{opt}	Optimum avalanche gain
η	Quantum efficiency
n	Refractive index
n_{eff}	Effective refractive index
n_i	Total intrinsic carrier density
n_n	Electron density in n-type material
n_p	Electron density in p-type material
$<n^2>_T$	Total m.s. output noise voltage
N_a	Acceptor atom density
N_c	Density of electrons in conduction band
N_d	Donor atom density
N_v	Density of holes in the valence band
μ_o	Permeability of free space ($4\pi \times 10^{-7}$ H/m)
μ_r	Relative permeability
N	Mode number (integer)
N_g	Group refractive index
N_{max}	Maximum number of modes
NA	Numerical aperture
p_n	Hole density in n-type material
p_p	Hole density in p-type material
P	Average received power
P_e	Probability of error
q	Electronic charge (1.6×10^{-19} C)
Q	Signal-to-noise parameter
r_π	Base-emitter resistance
$r_{bb'}$	Base-spreading resistance
r_e	Reflection coefficient
R_1, R_2	Mirror reflectivity in resonator
R_b	Photodiode load resistor
R_f	Feedback resistor
R_{in}	Preamplifier input resistance
R_j	Photodiode shunt resistance
R_L	Load resistor
R_o	Photodiode responsivity (A/W)

R_s	Photodiode series resistance
R_T	Low-frequency transimpedance
σ	r.m.s. width of Gaussian distribution
σ_{mat}	Material dispersion per unit length
σ_{mod}	Modal dispersion per unit length
σ_{off}	r.m.s. output noise voltage for logic 0
σ_{on}	r.m.s. output noise voltage for logic 1
σ_{wg}	Waveguide dispersion per unit length
S	Instantaneous power flow (Poynting vector)
S_{av}	Average power flow
S_E	Series noise generator (V^2/Hz)
$S_{eq}(f)$	Equivalent input noise current spectral density (A^2/Hz)
S_I	Shunt noise generator (A^2/Hz)
S/N	Signal-to-noise ratio
τ	Time constant
τ_n	Electron lifetime in p-type material
τ_{nr}	Non-radiative recombination time
τ_p	Hole lifetime in n-type material
τ_{ph}	Stimulated photon lifetime
τ_r	Radiative recombination time
τ_{sp}	Spontaneous photon lifetime
t_e	Transmission coefficient
T	Absolute temperature (K)
v_g	Group velocity
v_{max}	Maximum output signal voltage
v_{min}	Minimum output signal voltage
v_p	Phase velocity
V	Normalised frequency in a waveguide
V_{br}	Reverse breakdown voltage
V_s	Output signal voltage
V_T	Threshold voltage
y	Normalised frequency variable
Z	Impedance of dielectric to TEM waves
$Z_c(s)$	Closed-loop transimpedance
$Z_f(s)$	Feedback network transimpedance
Z_{in}	Total input impedance
Z_o	Impedance of free space
$Z_o(s)$	Open-loop transimpedance
$Z_T(\omega)$	Transimpedance
θ_i	Angle of incidence
θ_r	Angle of reflection
θ_t	Angle of transmission

About the Author

Martin Sibley graduated with a B.Sc. (Hons) degree in Electrical Engineering from the Polytechnic of Huddersfield in 1981. He then stayed on to study for a Ph.D. in preamplifier design for optical receivers working in conjunction with British Telecom Research Labs (BTRL). A Post-Doctoral Research Fellowship (1984–1986) funded by BTRL found him designing a Digital PPM coder/decoder for use at 300 Mbit/s. He then joined the academic staff at the Polytechnic of Huddersfield (now the University of Huddersfield) where he carried on researching PPM systems and free-space optical communications. He is the author of two other textbooks on electromagnetism and communications. He is now retired and lives in South Wales where he continues to write.

Chapter 1
Introduction

1.1 Historical Background

The use of light as a means of communication is not a new idea; many civilisations used sunlight reflected off mirrors to send messages, and communication between warships at sea was achieved using Aldis lamps. Unfortunately, these early systems operated at very low data rates and failed to exploit the very large bandwidth of optical communication.

A glance at the electromagnetic spectrum shown in Fig. 1.1 reveals that visible light extends from 0.4 to 0.7 μm which converts to a bandwidth of 320 THz (1 THz $= 10^{12}$ Hz). Even if only 1 per cent of this capability were available, it would still allow for 80 billion, 4 kHz voice channels! (If we could transmit these channels by radio, they would occupy the whole of the spectrum from d.c. right up to the far infrared. As well as not allowing for any radio or television broadcasts, the propagation characteristics of the transmission scheme would vary tremendously.) The early optical systems used incandescent white light sources, the output of which was interrupted by a hand-operated shutter. Apart from the obvious disadvantage of a low transmission speed, a white light source transmits all the visible, and some invisible, wavelengths at once. If we draw a parallel with radio systems, this is equivalent to a radio transmitter broadcasting a single programme over the whole of the radio spectrum – very inefficient! Clearly, the optical equivalent of an oscillator was needed before light-wave communications could develop.

A breakthrough occurred in 1960, with the invention of the ruby laser by T. H. Maimon [1], working at Hughes Laboratories, USA. For the first time, an intense, coherent light source operating at just one wavelength was made available. It was this development that started a flurry of research activity into optical communications.

Early experiments were carried out with line-of-sight links; however, it soon became apparent that some form of optical waveguide was required. This was because too many things can interfere with light-wave propagation in the

© Springer Nature Switzerland AG 2020
M. Sibley, *Optical Communications*, https://doi.org/10.1007/978-3-030-34359-0_1

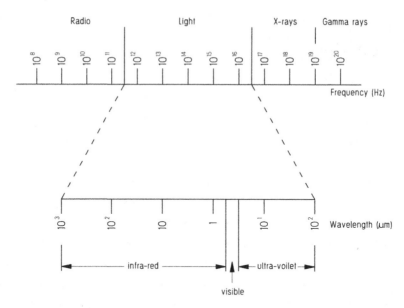

Fig. 1.1 The electromagnetic spectrum

atmosphere: fog, rain, clouds, and even the occasional flock of pigeons. Hollow metallic waveguides were initially considered, but, because of their impracticality, they were soon ruled out. By 1963, bundles of several hundred glass fibres were already being used for small-scale illumination. However, these early fibres had very high attenuations (>1000 dB/km), and so their use as a transmission medium for optical communications was not considered.

It was in 1966 that C. K. Kao and G. A. Hockman [2] (working at the Standard Telecommunications Laboratories, UK) postulated the use of glass fibres as optical communications waveguides. The idea was initially treated with some scepticism because of the high attenuation of the glass; in order to compete with existing coaxial cable transmission lines, the glass fibre attenuation had to be reduced to less than 20 dB/km. However, Kao and Hockman studied the loss mechanisms, and, in 1970, workers at the Corning Glass Works, USA, produced a fibre with the required attenuation. This development led to the first laboratory demonstrations of optical communications with glass fibre, in the early 1970s [3]. A study of the spectral response of glass fibres showed the presence of low-loss transmission windows at 850 nm, 1.3 µm and 1.55 µm. Although the early optical links used the 850 nm window, the longer wavelength windows exhibit lower losses, typically 0.2 dB/km, and so most modern links use 1.3 and 1.55 µm wavelength light.

While work progressed on reducing fibre attenuation, laser development continued apace. Ruby lasers have to be 'pumped' with the light from a flash lamp, and so the modulation speed is very low. The advent of the semiconductor laser, in 1962, meant that a fast light source was available. The material used was gallium-arsenide, *GaAs*, which emits light at a wavelength of 870 nm. With the discovery of the

850 nm window, the wavelength of emission was reduced by doping the GaAs with aluminium, *Al*. Later modifications included different laser structures to increase device efficiency and lifetime. Various materials were also investigated, to produce devices for operation at 1.3 and 1.55 μm. Unfortunately, lasers are quite expensive, and so low-cost light-emitting diodes, *LEDs*, have also been developed. Semiconductor sources are now available which emit at any one of many wavelengths, with modulation speeds of several tens Gbit/s being routinely achieved.

At the receiver, a photodetector converts the optical signal back into an electrical one. The early optical links used avalanche photodiodes, *APDs*, which exhibit current multiplication, that is, the single electron-hole pair produced by the detection of a photon of light generates more electron-hole pairs, so amplifying the signal. In 1973, S. D. Personick [4] (working at Bell Laboratories in the USA) analysed the performance of an optical PCM receiver. This theoretical study showed that an APD feeding a high input impedance preamplifier, employing an FET input stage, would result in the best receiver sensitivity. Unfortunately, the early APDs required high bias voltages, typically 200–400 V, and this made them unattractive for use in terminal equipment.

It was in 1978 that D. R. Smith, R. C. Hooper and I. Garrett [5] (all working at British Telecom Research Laboratories, Martlesham Heath, UK) published a comparison between an APD and a PIN photodiode followed by a low-capacitance, microwave FET input preamplifier (the so-called *PINFET* receiver). They showed that PINFET receivers using a hybrid thick-film construction technique could achieve a sensitivity comparable to that of an APD receiver. They also indicated that PIN receivers for the 1.3 and 1.55 μm transmission windows would outperform an equivalent APD receiver. (The reasons for this will become clear when we discuss photodiodes in Chap. 4). So, the use of PINFET receivers operating in the long-wavelength transmission windows meant that signals could be sent over very long distances – ideal for trunk route telephone links.

The work on long-haul routes aided the development of short-haul industrial links. From an industrial viewpoint, the major advantage of an optical link is its immunity to electromagnetic interference. Hence optical fibre links can operate in electrically noisy environments which would disrupt a hard-wire system. For short-haul applications, expensive low-loss glass fibres, lasers and very sensitive receivers are not required. Instead, all-plastic fibres, LEDs and low-cost bipolar preamplifiers are often used. These components are readily available on the commercial market and are usually supplied with connectors attached for ease of use.

In long-haul links, regenerators were needed to combat the effects of pulse dispersion and attenuation. Such regenerators consisted of a detector and amplifier, timing extraction, a decision gate and a light source and were needed every 30–40 km, clearly an inconvenience in transoceanic links. The development of erbium-doped fibre amplifiers (*EDFAs*) meant that signals could be amplified without being detected first. The only drawback was the build-up of pulse dispersion, but the use of dispersion-shifted fibre operating at a wavelength of 1.55 μm reduced the effects and the losses.

Early wavelength division multiplexing (*WDM*) systems used the 1.3 μm and 1.55 μm fibre transmission windows. This was because sources were available at these wavelengths although they had quite large spectral spreads (*linewidth*). Lasers have subsequently been developed that produce a single frequency with a linewidth of sub-MHz. This means that a particular data stream of typically 40 Gbit/s can be allocated to an individual wavelength in the same way that a radio station is allocated a particular frequency. We now have commercially available fibre links with an aggregate data rate in excess of 1 Tbit/s with Pbit/s links in the lab. Such links use spatial division multiplexing (*SDM*) in which 12 cores share a common cladding. (A 1 Pbit/s link is capable of transmitting 5000 HDTV signals in 1 second!) Computer communications benefitted from this research and development, and we now have 100 Gbit/s Ethernet links. Such high bit rates are needed to carry the data produced by our information society.

1.2   The Optical Communications Link

An optical communications link is basically the same as any other communications link: there is a source, a channel (either optical fibre or air) and a receiver (Fig. 1.2). This is the same as a more familiar radio link. However, there are some obvious differences in configuration. The source is either a laser or a light-emitting diode (*LED*). The channel is optical fibre (or free-space), which serves to guide the light, and the detector is a photodiode that converts the light back into an electrical signal. For the fibre to guide the light, it must have a *core* surrounded by a *cladding* of lower refractive index so that total internal reflection occurs. Some fibre types have a relatively wide core of 50 μm diameter; others have fibre dimensions that are very small with the core being 5–8 μm in diameter and the cladding 125 μm in diameter. We can study the propagation of light using ray-path analysis and that will introduce us to some useful concepts. However, we will have to use Maxwell's equations to study propagation of light in optical fibre in detail. This is covered in Chap. 2.

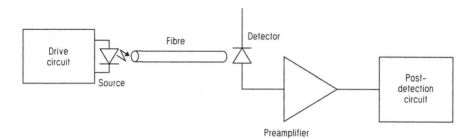

Fig. 1.2 A basic optical communications link

Table 1.1 The SDH standard

SDH level	Bit rate (Mbit/s)	Channels	Pulse time
STM-1	155.52	2016	6.43 ns
STM-4	622.08	8064	1.61 ns
STM-16	2488.32	32,256	0.40 ns
STM-64	9953.28	129,024	100 ps
STM-256	39,813.12	516,096	25 ps

The transmitter is a laser diode or an LED. We are familiar with LEDs as displays, but they can be produced to emit light in the infrared which is needed for optical fibre communications. LEDs can also be used for illumination – visible light – and it is possible to modulate them. The result is a form of free-space optical communication called visible light communication (VLC). Lasers are widely used in optical fibre communication and free-space optical communications (FSOC). There are currently many different types of laser depending on the application; sources are considered in Chap. 3.

Chapter 4 examines detectors – *PIN* and avalanche photodiodes (*APDs*). A photodetector should respond to the wavelength of operation, be fast enough to detect the signal and be quiet in terms of noise. The power level at the end of an optical link can be as low as 1 µW, and so the specification of the photodetector is very important.

Chapter 5 deals with the theory underpinning the performance of optical receivers. Given a particular receiver system (photodiode, transimpedance preamplifier, post-detection filter), it is possible to predict the sensitivity (the optical power required for a certain performance criterion). Chapter 6 discusses the design of preamplifiers for use in optical receivers. Chapter 7 brings it all together by discussing several optical links. It is always difficult to predict the future in such a fast-moving field, but this chapter also outlines some of the new innovations and what part they might play.

Before we move on, it will be instructive to examine the standardised bit rates used internationally. Table 1.1 lists the bit rate for the synchronous digital hierarchy (*SDH*) standard. As can be seen, the data rate increases by a factor of 4 for every layer. (*STM* stands for synchronous transport module.) It might be thought that the number of channels should be higher; after all, 155.52 Mbit/s divided by 64 kbit/s gives 2 430 channels rather than 2 016. The difference is due to there being an overhead used for control purposes.

STM-256 links are commercially available now, and higher speed links have been developed in the laboratory. The pulse time quoted in the table is the length of time taken to transmit a pulse using the non-return-to-zero (*NRZ*) format (Fig. 1.3). This is the preferred format as it has a low spectral width (dealt with in Chap. 5).

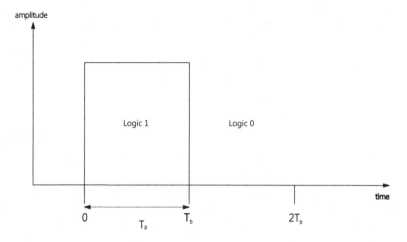

Fig. 1.3 The NRZ pulse format

Problems

1. An audio signal is sampled at a rate of 8 kHz by an 8 bit analogue to digital convertor (ADC). Determine the bit rate at the output of the ADC.

   [64 kbit/s]

2. Determine the number of speech channels that can be carried by a single optical fibre link capable of transmitting 1 Gbit/s, 10 Gbit/, 100 Gbit/s, 1 Tbit/s and 10 Tbit/s. Compare to the population of some countries. (In reality there is an overhead due to synchronising and so these are estimates.)

   [15.625 k, 156.25 k, 1.5625 M, 15.625 M, 156.25 M – the number of phone calls is similar to the population of some countries]

3. Convert wavelengths of 0.7 μm and 1.55 μm to frequencies. What is the bandwidth and what is the available bandwidth assuming 10% usage?

   [428 THz, 194 THz, 234 THz, 23.4 THz]

4. The loss of an optical fibre had to be lower than 20 dB/km in order to compete with coaxial cable. Determine this attenuation as a ratio and the received power after 1 km if the launch power is 3 mW.

   [100, 30 μW]

5. A 1.55 μm laser has a linewidth of 1 nm. Determine the spread of frequencies of this laser. What is the linewidth expressed as a percentage of the carrier wavelength?

   [125 GHz, 0.064%]

Recommended Reading

1. Maimon TH (1960) Stimulated optical radiation in ruby. Nature 187:493–494
2. Kao CK, Hockman GA (1966) Dielectric-fibre surface waveguides for optical frequencies. Proc IEEE 113:1151–1158
3. Kaiser P, Astle HW (1974) Low-loss single-material fibers made from pure fused silica. Bell Syst Tech J 53:1021–1039
4. Personick SD (1973) Receiver design for digital fiber optic communication systems, parts I and II. Bell Syst Tech J 52:843–886
5. Smith DR, Hooper RC, Garrett I (1978) Receivers for optical communications: a comparison of avalanche photodiodes with PIN-FET hybrids. Opt Quant Electron 10:293–300
6. Fleisch D (2008) A student's guide to Maxwell's equations. Cambridge University Press, UK, Cambridge

Chapter 2
Optical Fibre

In most optical communication links, it is the optical fibre that provides the transmission channel. The fibre consists of a solid cylinder of transparent material, the *core*, surrounded by a *cladding* of similar material. Light waves propagate down the core in a series of plane wavefronts, or *modes*; the simple light ray path used in elementary optics is an example of a mode. For this propagation to occur, the refractive index of the core must be larger than that of the cladding, and there are two basic structures which have this property: *step-index* and *graded-index* fibres. Of the step-index types, there are multimode, *MM*, fibres (which allow a great many modes to propagate) and single-mode, *SM*, fibres (which only allow one mode to propagate). Although graded-index fibres are normally MM, some SM fibres are available. The three fibre types, together with their respective refractive index profiles, are shown in Fig. 2.1. (In this figure, *n* is the refractive index of the material.) The cross-hatched area represents the cladding, the diameter of which ranges from 125 µm to a typical maximum of I mm. The core diameter can range from 5 µm, for SM fibres, up to typically 50 µm for large-core MM fibres.

Most of the optical fibres in use today are made of either silica glass ($SiO2$) or plastic. The change in refractive index, between the core and cladding, is achieved by the addition of certain dopants to the glass; all-plastic fibres use different plastics for the core and cladding. All-glass, SM fibres exhibit very low losses and high bandwidths, which make them ideal for use in long-haul telecommunications routes. Unfortunately, such fibres are expensive to produce and so are seldom found in short-haul (less than 500 m length) industrial links.

Large-core fibres for use in medical and industrial applications are generally made of plastic, making them more robust than the all-glass types and much cheaper to manufacture. However, the very high attenuation and low bandwidth of these fibres tend to limit their uses in communications links. Medium-haul routes, between 500 m and 1 km lengths, generally use plastic cladding/glass core fibre, otherwise known as *plastic clad silica*, or *PCS*. All-plastic and PCS fibres are almost exclusively step-index, multimode types.

M. Sibley, *Optical Communications*, https://doi.org/10.1007/978-3-030-34359-0_2

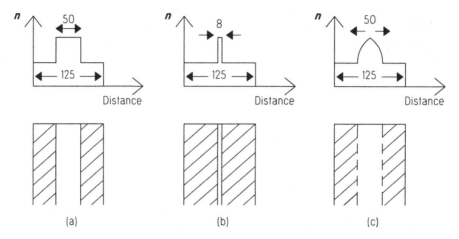

Fig. 2.1 Typical refractive index profiles of (**a**) step-index multimode, (**b**) step-index single-mode and (**c**) graded-index multimode fibres. (All dimensions are in μm)

The attenuation and bandwidth of an optical fibre will determine the maximum distance that signals can be sent. Attenuation is usually expressed in dB/km, while bandwidth is usually quoted in terms of the *bandwidth-length product*, which has units of GHz km or MHz km. Attenuation depends on impurities in the core, and so the fibre must be made from very pure materials. To some extent the bandwidth also depends on the core impurities; however, the bandwidth is usually limited by the number of propagating modes. This explains why single-mode fibres have a very large bandwidth.

In this chapter we shall examine the properties and design of optical fibres. Initially we will solve Maxwell's equations in an infinite block of dielectric material (glass) and then consider propagation in a planar dielectric waveguide. When we come to examine propagation in optical fibre, we will solve Maxwell's equations as applied to a cylindrical waveguide. This involves some rather complicated mathematics which some readers may prefer to omit at a first reading. In view of this, important results from the full analysis are quoted in the relevant sections. (The work in this chapter assumes that the reader is familiar with Maxwell's equations. Most books on electromagnetism cover the derivation of these equations.)

2.1 Propagation of Light in a Dielectric

In some instances it is convenient to treat light as a stream of particles, or *photons*, and in others as an electromagnetic wave. Here we will treat light as a wave and apply Maxwell's equations to study light-wave propagation. We will consider a plane wavefront, of arbitrary optical frequency, travelling in an infinite block of dielectric (glass). This will give us a valuable insight into certain fibre characteristics which we cannot easily explain in terms of simple geometric ray optics.

2.1.1 The Wave Equation

In order to study the variation of the **E** and **H** fields in a dielectric, we need to derive the relevant wave equations. We take as our starting point the following Maxwell's equations:

$$\nabla \times \mathbf{E} = -\mu \frac{\partial \mathbf{H}}{\partial t} \tag{2.1a}$$

and

$$\nabla \times \mathbf{H} = \varepsilon \frac{\partial \mathbf{E}}{\partial t} + \sigma \mathbf{E} \tag{2.1b}$$

If **E** and **H** vary sinusoidally with time at the frequency of the light we are transmitting, then we can use the phasor forms of **E** and **H**. Thus we can write

$$\mathbf{E} = E_x \exp\left(j\omega t\right)\mathbf{a}_x \quad \text{and} \quad \mathbf{H} = H_y \exp\left(j\omega t\right)\mathbf{a}_y$$

where $\mathbf{a}_x$ and $\mathbf{a}_y$ are the x- and y-axes unit vectors. We can now write (2.1a) and (2.1b) as

$$\frac{\partial \mathbf{E}}{\partial z} = -j\omega\mu\mathbf{H} \tag{2.2a}$$

and

$$-\frac{\partial \mathbf{H}}{\partial z} = j\omega\varepsilon\mathbf{E} + \sigma\mathbf{E} \tag{2.2b}$$

We can manipulate these two equations to give the wave equations which describe the propagation of a plane transverse electromagnetic (TEM) wave in the material. Thus, if we differentiate (2.2a) with respect to z, and substitute from (2.2b), we get

$$\frac{\partial^2 \mathbf{E}}{\partial z^2} = -\omega^2\mu\varepsilon\mathbf{E} + j\omega\mu\sigma\mathbf{E} \tag{2.3}$$

and, if we differentiate (2.2b) with respect to z, and substitute from (2.2a), we get

$$\frac{\partial^2 \mathbf{H}}{\partial z^2} = -\omega^2 \mu \varepsilon \mathbf{H} + j\omega\mu\sigma \mathbf{H} \tag{2.4}$$

If we now let $\gamma^2 = -\omega^2 \mu \varepsilon + j\omega\mu\sigma$, one possible solution to these equations is

$$\mathbf{E} = E_{xo} \exp (j\omega t) \exp (-\gamma z) \mathbf{a}_x \tag{2.5}$$

and

$$\mathbf{H} = H_{yo} \exp (j\omega t) \exp (-\gamma z) \mathbf{a}_y \tag{2.6}$$

where the subscript o denotes the values of E and H at the origin of a right-handed Cartesian coordinate set and γ is known as the *propagation coefficient*. Writing $\gamma = \alpha + j\beta$, where α and β are the *attenuation* and *phase coefficients*, respectively, we get

$$\mathbf{E} = E_{xo} e(-\alpha z) \cos (\omega t - \beta z) \mathbf{a}_x \tag{2.7}$$

and

$$\mathbf{H} = H_{yo} e(-\alpha z) \cos (\omega t - \beta z) \mathbf{a}_y \tag{2.8}$$

These equations describe a TEM wave travelling in the positive z-direction, undergoing attenuation as $\exp(-\alpha z)$. The E and H fields are orthogonal to each other and, as Fig. 2.2 shows, perpendicular to the direction of propagation.

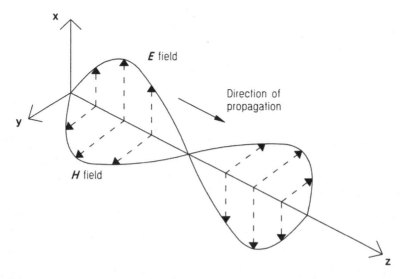

Fig. 2.2 Variation of E and H for a TEM wave propagating in the z-direction

2.1.2 Propagation Parameters

Equations (2.7) and (2.8) form the starting point for a more detailed study of light-wave propagation. However, before we proceed much further, we will find it useful to derive some propagation parameters. From the previous section, the propagation coefficient, γ, is given by

$$\gamma = \alpha + j\beta \quad \text{and} \quad \gamma^2 = -\omega^2 \mu \varepsilon + j\omega\mu\sigma$$

Hence it is a simple matter to show that

$$\alpha^2 - \beta^2 = -\omega^2 \mu \varepsilon \tag{2.9}$$

and

$$2\alpha\beta = \omega\sigma\mu \tag{2.10}$$

As glass is an insulator, the conductivity is very low, $\sigma = 0$, and the relative permeability is approximately unity, $\mu_r = 1$. Over a distance of a few wavelengths, this results in $\alpha = 0$ (which implies zero attenuation), and so we can write the $\mathbf{E}$ and $\mathbf{H}$ fields as

$$\mathbf{E} = E_{xo} \cos(\omega t - \beta z)\mathbf{a}_x \tag{2.11}$$

and

$$\mathbf{H} = H_{yo} \cos(\omega t - \beta z)\mathbf{a}_y \tag{2.12}$$

where $\beta = \omega\sqrt{\mu_o \varepsilon}$. We can study the propagation of these fields by considering a point on the travelling wave as t and z change.

Figure 2.3 shows the sinusoidal variation of the E field with time and distance. If we consider the point A then, at time $t = 0$ and distance $z = 0$, the amplitude of the wave is zero. At time $t = t_1$, the point A has moved a distance z_1, and, as the amplitude of the wave is still zero, we can write

$$\sin(\omega t_1 - \beta z_1) = 0 \quad \text{and so} \quad \omega t_1 = \beta z_1 \tag{2.13}$$

Thus we can see that the constant phase point, A, propagates along the z-axis at a certain velocity. This is the *phase velocity*, v_p, given by

$$v_p = \frac{z_1}{t_1} = \frac{\omega}{\beta} = \frac{\omega}{\omega\sqrt{\mu_o \varepsilon}} = \frac{1}{\sqrt{\mu_o \varepsilon}} \tag{2.14}$$

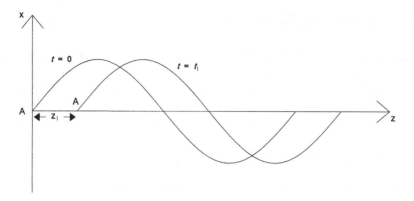

Fig. 2.3 Illustrative of the phase velocity of a constant phase point, A, on the E field of a TEM wave

If the dielectric is free space, then v_p is 3×10^8 m/s (the velocity of light in a vacuum, c). This leads us directly on to the definition of refractive index. For the dielectric, $\varepsilon = \varepsilon_o \, \varepsilon_r$ and so

$$v_p = \frac{1}{\sqrt{\mu_o \varepsilon_o}} \times \frac{1}{\sqrt{\varepsilon_r}} = \frac{c}{n} \tag{2.15}$$

where n is the *refractive index* of the dielectric. It is a simple matter to show that the wavelength of the light in the dielectric, λ, is

$$\lambda = \frac{2\pi}{\beta} = \frac{2\pi v_p}{\omega} = \frac{2\pi c}{n\omega} = \frac{\lambda_o}{n} \tag{2.16}$$

where λ_o is the wavelength of the light in free space. This leads us to an alternative definition for β:

$$\beta = \frac{2\pi n}{\lambda_o} = nk_o \tag{2.17}$$

where k_o is the *free-space propagation constant*. (It is interesting to note that if ε_r and hence n vary with frequency, then β and v_p will also vary. Thus, if we have two light waves of slightly different frequencies, the two waves will travel at different velocities, and the signal is said to be *dispersed*. We shall return to this point in the next section.)

The impedance of the dielectric to TEM waves, Z, equals E/H, and we can find E as a function of H by using (2.2a). Thus

$$\frac{\partial \mathbf{E}}{\partial z} = -j\mu_0\omega\mathbf{H} \quad \text{becomes} \quad -j\beta\mathbf{E} = -j\mu_0\omega\mathbf{H} \quad \text{and so}$$

$$Z = \frac{\mu_0\omega}{\beta} = \sqrt{\frac{\mu_0}{\varepsilon_0\varepsilon_r}} = \frac{Z_0}{n} \tag{2.18}$$

where Z_0 is the impedance of free space, 377 Ω.

One final useful parameter is the power flow. If we take the cross product of $\mathbf{E}$ and $\mathbf{H}$, we will get a third vector, acting in the direction of propagation, with units of W/m^2, that is, power flow per unit area. This vector is known as the *Poynting vector*, S, and its *instantaneous* value is given by

$$S = \mathbf{E} \times \mathbf{H} \tag{2.19}$$

We can find the *average* power flow, S_{av}, in the usual way by integrating (2.19) over one period and then dividing by the period. In phasor notation form, S_{av} will be given by

$$S_{av} = R_e\left\{\frac{1}{2}\mathbf{E} \times \mathbf{H}^*\right\} \tag{2.20}$$

where R_e denotes 'the real part of . . . ', $\mathbf{H}^* = H\exp(-j[\omega t - \varphi])$ and φ is the temporal phase angle between the E and H fields. (In geometric optics, ray paths are drawn in the direction of propagation and normal to the plane of $\mathbf{E}$ and $\mathbf{H}$. Thus the ray path has the direction of power flow.)

Example
Light of wavelength 600 nm is propagating in a block of transparent material which has the following characteristics:

$$\mu_r = 1; \varepsilon_r = 5; \sigma = 3 \times 10^{-4}\,\text{S/m}$$

Determine the following parameters:

(a) **Attenuation and phase coefficients**
(b) **Phase velocity**
(c) **Refractive index**
(d) **Impedance to TEM waves**

If the light has an electric field strength of 5 kV/m, determine the magnetic field strength and the average power per unit area.

(continued)

(a) We can find the attenuation and phase coefficients from Eqs. (2.9) and (2.10). So

$$\alpha^2 - \beta^2 = -\omega^2 \mu \varepsilon$$

and

$$2\alpha\beta = \omega\sigma\mu$$

Thus

$$\alpha^2 - \beta^2 = -\left[\frac{2\pi c}{\lambda}\right]^2 \mu_0 \varepsilon_0 \varepsilon_r$$
$$= -5.5 \times 10^{14}$$

and

$$2\alpha\beta = 1.2 \times 10^6$$

Hence

$$\alpha^2 - \left[\frac{1.2 \times 10^6}{2\alpha}\right]^2 = -5.5 \times 10^{14}$$

and so $\alpha = 0$ or imaginary. Thus the light experiences negligible attenuation. By following a similar procedure, we find that $\beta = 2.34 \times 10^7$ rad/m. This is quite large but it must be remembered that the wavelength of the light is small.

(b) The phase velocity is given by $v_p = \omega/\beta$. Thus

$$v_p = \frac{3.14 \times 10^{15}}{2.34 \times 10^7} = 1.34 \times 10^8 \, \text{m/s}$$

(c) The refractive index is given by

$$n = \frac{c}{v_p} = 2.24$$

or

(continued)

$$n = \sqrt{\varepsilon_r}$$

or

$$n = \frac{\beta}{k_o}$$

(d) The impedance of the material is

$$Z = \frac{Z_0}{n} = \frac{377}{2.24} = 168\ \Omega$$

Now, the magnitude of the E field is 5 kV/m. As $Z = E/H$, we can write

$$H = \frac{E}{Z} = \frac{5 \times 10^3}{168}\ \text{A/m}$$

We can find the average power per m^2 by using

$$S_{av} = \frac{1}{2}E \times H = 74.5\,\text{kW/m}^2$$

This represents a power of 1.86 W in a 25 mm^2 area.

2.1.3 Group Velocity and Material Dispersion

As we have seen, the velocity of light in a dielectric depends upon the refractive index. However, because of the atomic interactions between the material and the optical signal, refractive index varies with wavelength and so any light consisting of several different wavelengths will be dispersed. (A familiar example of dispersion is the spectrum produced when white light passes through a glass prism.) To examine the effect of dispersion on an optical communication link, we will consider intensity, or amplitude, modulation of an optical signal.

If a light source of frequency ω_c is amplitude modulated by a single frequency, ω_m, then the electric field intensity, e_{AM}, at a certain point in space will be

$$\begin{aligned} e_{AM} &= E_{xo}(1 + m \cos \omega_m t) \cos \omega_c t \\ &= E_{xo}\left(\cos \omega_c t + \frac{m}{2}[\cos(\omega_c + \omega_m)t + \cos(\omega_c - \omega_m)t] \right) \end{aligned} \tag{2.21}$$

where m is the depth of modulation. Thus there are three individual frequency components: the carrier signal, ω_c; the upper-side frequency, $\omega_c + \omega_m$; and the

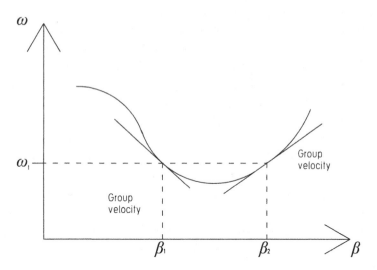

Fig. 2.4 Showing the relationship between the phase and group velocity of a modulated TEM wave

lower-side frequency, $\omega_c - \omega_m$. As $\beta = 2\pi/\lambda$, each of these components will have their own value of β. So, the variation of e_{AM} with distance, z, can be written as

$$e_{AM} = E_{xo} \cos{(\omega_c t - \beta z)} + E_{xo}\frac{m}{2}\cos{[(\omega_c + \delta\omega)t - (\beta + \delta\beta)z]}$$
$$+ E_{xo}\frac{m}{2}\cos{[(\omega_c - \delta\omega)t - (\beta - \delta\beta)z]} \tag{2.22}$$

where $\delta\omega$ has replaced ω_m, and we have assumed that the variation of β with ω is linear around ω_c (Fig. 2.4).

The last two terms in (2.22) – the two side frequencies – can be written as

$$E_{xo}\frac{m}{2}\cos{[(\omega_c t - \beta z) + (\delta\omega t - \delta\beta z)]} + E_{xo}\frac{m}{2}\cos{[(\omega_c t - \beta z) - (\delta\omega t - \delta\beta z)]}$$
$$= E_{xo}m\cos{(\omega_c t - \beta z)}\cos{(\delta\omega t - \delta\beta z)}$$

$$\tag{2.23}$$

and so we can describe the total wave by

$$E_{xo}\cos{(\omega_c t - \beta z)} + E_{xo}m\cos{(\omega_c t - \beta z)}\cos{(\delta\omega t - \delta\beta z)} \tag{2.24}$$

Examination of this equation shows that the first term, the carrier wave, propagates at the familiar phase velocity. However, the second term, the modulation envelope, travels at a velocity of $\delta\omega/\delta\beta$ known as the *group velocity*, v_g. Thus we can write

$$v_p = \frac{\omega}{\beta} \qquad (2.25)$$

and

$$v_g = \frac{\delta\omega}{\delta\beta} \qquad (2.26)$$

From these equations it should be evident that v_g is the gradient of a graph of ω against β, as in Fig. 2.4. Examination of Fig. 2.4 shows that the group velocity is dependent on frequency. Thus different frequency components in a signal will travel at different group velocities, and so will arrive at their destination at different times. For digital modulation of the carrier, this results in smearing, or *dispersion*, of the pulses, which affects the maximum rate of modulation. The variation of refractive index with frequency is dependent on the glass material, and so this form of dispersion is known as *material dispersion*.

To observe the effect of material dispersion, let us derive the difference in propagation times, $\delta\tau$, for the two sidebands previously considered. We can express $\delta\tau$ as

$$\delta\tau = \frac{d\tau}{d\lambda}\delta\lambda \qquad (2.27)$$

where $\delta\lambda$ is the wavelength difference between the lower and upper sideband and $d\tau/d\lambda$ is known as the material dispersion coefficient, D_{mat}. If we consider a unit length, then $\tau = 1/v_g$, and so

$$D_{\text{mat}} = \frac{d\tau}{d\lambda} = \frac{d}{d\lambda} \times \frac{1}{v_g} \qquad (2.28)$$

Now,

$$\frac{1}{v_g} = \frac{d\beta}{d\omega} = \frac{d\lambda}{d\omega} \times \frac{d\beta}{d\lambda} = -\frac{\lambda_0^2}{2\pi c} \times \frac{d\beta}{d\lambda} = -\frac{\lambda_0^2}{2\pi c} \times \frac{d}{d\lambda} k_0 n$$

$$= -\frac{\lambda_0^2}{2\pi c} \times \frac{d}{d\lambda} \frac{2\pi n}{\lambda_0} = -\frac{\lambda_0^2}{c} \times \frac{d}{d\lambda} \times \frac{n}{\lambda_0} = \frac{1}{c}\left[n - \lambda_0 \frac{dn}{d\lambda}\right] = \frac{N_g}{c} \qquad (2.29)$$

where N_g is the *group refractive index* – compare with the definition of refractive index given by Eq. (2.15). So,

$$D_{mat} = \frac{d\tau}{d\lambda} = \frac{d}{d\lambda} \times \frac{N_g}{c} = \frac{1}{c}\left[\frac{dn}{d\lambda} - \frac{\lambda_0 d^2 n}{d\lambda^2} - \frac{dn}{d\lambda}\right] = -\frac{\lambda_0}{c}\frac{d^2 n}{d\lambda^2} \qquad (2.30)$$

(Use has been made of (2.16).) The negative sign in (2.30) shows that the upper sideband signal, the lowest wavelength, arrives before the lower sideband, the highest wavelength. The units of D_{mat} are normally ns/nm/km (remember that we are considering a unit length of material). So, in order to find the dispersion in ns, we need to multiply D_{mat} by the wavelength difference between the minimum and maximum spectral components and the length of the optical link. As the link length is variable, the material dispersion is usually expressed in units of time per unit length – symbol σ_{mat}.

Before we consider an example, it is worth noting that if the group velocity is the same as the phase velocity, as with air, then the material will be dispersionless. We should also note that, when we consider the planar waveguide, we should resolve the material dispersion along the horizontal axis. However, with multimode waveguides, modal dispersion is generally more significant than the material dispersion.

Example
A 100 MHz sinewave causes amplitude modulation of a 600 nm wavelength light source. The resultant light propagates through a dispersive medium with $D_{mat} = 50$ ps/nm/km. Determine the material dispersion.

As the light source is amplitude modulated, the modulated light will consist of the carrier wave and two sidebands spaced 100 MHz either side of the carrier. Thus the spread in wavelength is

$$\delta\lambda = 2.4 \times 10^{-4}\, nm$$

and so the material dispersion is

$$\sigma_{mat} = 0.012\, ps/km$$

This calculation assumes that the source is monochromatic, that is, the source generates light at a frequency 5×10^{14} Hz and *no other frequencies*. In practice, most light sources generate a range of wavelengths with the spread being known as the *linewidth* of the source. If we take a linewidth of 10 nm (i.e. the light consists of a range of frequencies from 4.96×10^{14} to 5.04×10 Hz), we find that

$$\delta\lambda = 10\, nm$$

and so

(continued)

$$\sigma_{mat} = 500\,ps/km$$

which is far higher than the dispersion introduced by the modulating signal alone.

From these calculations we can see that the spectral purity of the source can dominate σ_{mat} if the source linewidth is large. So, for high-data-rate or long-haul applications, it is important to use narrow linewidth sources (dealt with in the next chapter).

Figure 2.5 shows the variation of D_{mat} with λ for three typical *glass* fibres. Because D_{mat} passes through zero at wavelengths around 1.3 μm, which happens to coincide with one of the transmission windows, this was the most popular wavelength for long-haul links. This situation is now changing with the introduction of *dispersion-shifted* fibres, dealt with in Sect. 2.3.4, in which the zero dispersion point is at 1.55 μm – a transmission window which offers lower attenuation.

We have seen that the composition of the fibre causes dispersion of the signal due to the variation of group refractive index with wavelength. There are, however, three further forms of dispersion – *modal*, *waveguide* and *polarisation mode* – and we can examine the first two of these by considering propagation in a dielectric slab waveguide.

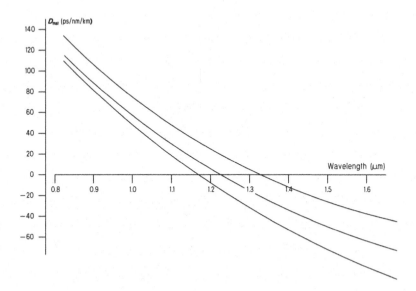

Fig. 2.5 Variation of material dispersion, D_{mat} with wavelength for three different glass fibres

2.2   Propagation in a Planar Dielectric Waveguide

In this section we shall consider propagation in a simple planar optical waveguide. In particular, we shall examine reflection and refraction of a light wave at the waveguide boundaries. This will lead to the conditions we must satisfy before successful propagation can occur and introduce us to modal and waveguide dispersion. Although the values of dispersion we will calculate will seem large, we should remember that *planar* optical waveguides are generally quite short in length, and so dispersion effects are usually insignificant. In spite of this, the work presented here will be useful when we consider optical fibre.

2.2.1   Reflection and Refraction at Boundaries

Figure 2.6 shows a transverse electric, *TE*, wave, $\mathbf{E_i}$ and $\mathbf{H_i}$, incident on a boundary between two dissimilar, non-conducting, dielectrics (the waveguide boundary). As can be seen, some of the wave undergoes reflection, $\mathbf{E_r}$ and $\mathbf{H_r}$, while the rest is transmitted (or *refracted*), $\mathbf{E_t}$ and $\mathbf{H_t}$. In order to determine the optical power in both waves, we can resolve the waves into their x, y and z components and then apply the boundary conditions.

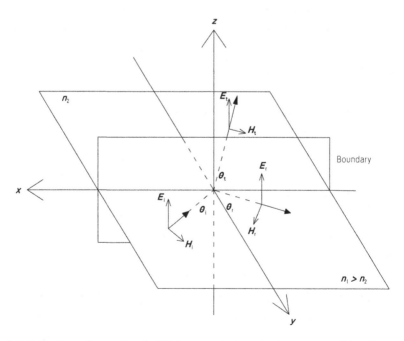

Fig. 2.6 Reflection and refraction of a TEM wave, at the boundary between two dielectric materials

Let us initially consider the E field as it crosses the boundary. We can express the incident field as the combination of a field propagating in the negative x-direction and another field travelling in the negative y-direction. Thus the propagation constant *associated with the propagating mode*, k, will be given by (using Pythagoras)

$$k^2 = \beta_x{}^2 + \beta_y{}^2$$

and so $\mathbf{E_i}$ can be written as

$$\mathbf{E_i} = \mathbf{a}_z E_o \exp\left(j\beta_1 [x\sin\theta_i + y\cos\theta_i] \right) \tag{2.31}$$

Similarly, we can write the reflected and transmitted fields as

$$\mathbf{E_r} = \mathbf{a}_z E_r \exp\left(j\beta_1 [x\sin\theta_r + y\cos\theta_r] \right) \tag{2.32}$$

$$\mathbf{E_t} = \mathbf{a}_z E_t \exp\left(j\beta_2 [x\sin\theta_t + y\cos\theta_t] \right) \tag{2.33}$$

(Here x and y refer to the *distances travelled along the respective axes*. This explains the absence of any negative signs in these equations.)

The boundary conditions at the interface require the tangential components of the E and H fields in both media to be continuous. If we initially consider the continuity of the E field, then, as these fields are already parallel to the interface, we can write

$$E_i + E_r = E_t \tag{2.34}$$

Dividing by E_i yields

$$1 + r_e = t_e \tag{2.35}$$

where r_e is the reflection coefficient, $r_e = E_r/E_i$, and t_e is the transmission coefficient, $t_e = E_t/E_i$. Thus we can write E_r and E_t as

$$\mathbf{E_r} = \mathbf{a}_z r_e E_o \exp\left(j\beta_1 [x\sin\theta_r + y\cos\theta_r] \right) \tag{2.36}$$

and

$$\mathbf{E_t} = \mathbf{a}_z t_e E_o \exp\left(j\beta_2 [x\sin\theta_t + y\cos\theta_t] \right) \tag{2.37}$$

We can substitute the expressions for the E fields at the interface, $y = 0$, back into (2.34) to give

$$E_o \exp\left(j\beta_1 x\sin\theta_i \right) + r_e E_o \exp\left(j\beta_1 x\sin\theta_r \right) = t_e E_o \exp\left(j\beta_2 x\sin\theta_t \right) \tag{2.38}$$

In order to satisfy (2.35), the exponential terms in (2.38) must be equal to each other, that is

$$\beta_1 \sin \theta_i = \beta_1 \sin \theta_r = \beta_2 \sin \theta_t$$

The first of these equalities yields

$$\theta_i = \theta_r \qquad (2.39)$$

which is *Snell's law of reflection*, that is, the angle of reflection equals the angle of incidence. The second equality gives

$$\sin \theta_t = \frac{\beta_1}{\beta_2} \sin \theta_i = \frac{k_o n_1}{k_o n_2} \sin \theta_i = \frac{n_1}{n_2} \sin \theta_i \qquad (2.40)$$

known as *Snell's law of refraction*, or simply Snell's law. (These equations should be familiar from geometric optics.)

In order to find an expression for r_e, let us now consider the second boundary relation – the continuity of the tangential H field. As the H fields act at right angles to the directions of propagation, we can write

$$\mathbf{H_i} = \left(-\mathbf{a}_x \cos \theta_i + \mathbf{a}_y \sin \theta_i \right) \frac{E_i}{Z_1} \qquad (2.41)$$

$$\mathbf{H_r} = \left(\mathbf{a}_x \cos \theta_r + \mathbf{a}_y \sin \theta_r \right) \frac{E_r}{Z_1} \qquad (2.42)$$

$$\mathbf{H_t} = \left(-\mathbf{a}_x \cos \theta_t + \mathbf{a}_y \sin \theta_t \right) \frac{E_t}{Z_2} \qquad (2.43)$$

Now, the tangential H field boundary relation gives, at $y = 0$,

$$\frac{E_o}{Z_1} \cos \theta_i \exp \left(j\beta_{1x}x \right) - r_e \frac{E_o}{Z_1} \cos \theta_r \exp \left(j\beta_{1x}x \right) = t_e \frac{E_o}{Z_2} \cos \theta_t \exp \left(j\beta_{2x}x \right) \quad (2.44)$$

where we have substituted for E_r and E_t. The new parameters β_{1x} and β_{2x} are the phase constants for media 1 and 2 *resolved onto the x-axis*, defined by

$$\beta_{1x} = \beta_1 \sin \theta_i = \beta_1 \sin \theta_r \qquad (2.45)$$

and

$$\beta_{2x} = \beta_2 \sin \theta_t \qquad (2.46)$$

As we have seen from (2.38), the exponential terms in (2.44) are all equal. Therefore (2.44) becomes

$$\frac{\cos \theta_i}{Z_1}(1 - r_e) = \frac{t_e \cos \theta_t}{Z_2} \tag{2.47}$$

Since $1 + r_e = t_e$, we can eliminate t_e from (2.47) to give

$$r_e = \frac{Z_2 \cos \theta_i - Z_1 \cos \theta_t}{Z_2 \cos \theta_i + Z_1 \cos \theta_t} = \frac{n_1 \cos \theta_i - n_2 \cos \theta_t}{n_1 \cos \theta_i + n_2 \cos \theta_t} \tag{2.48}$$

and, by using Snell's law, we can eliminate θ_i from (2.48) to give

$$r_e = \frac{\cos \theta_i - \sqrt{\left(\frac{n_2}{n_1}\right)^2 - \sin^2 \theta_i}}{\cos \theta_i + \sqrt{\left(\frac{n_2}{n_1}\right)^2 - \sin^2 \theta_i}} \tag{2.49}$$

Close examination of (2.49) reveals that r_e is unity if the term under the square root is zero, that is, $\sin^2 \theta_i = (n_2/n_1)^2$. Under these conditions, the reflected E field will equal the incident E field, and this is *total internal reflection*. The angle of incidence at which this occurs is the *critical angle*, θ_c, given by

$$\sin^2 \theta_c = \left[\frac{n_2}{n_1}\right]^2 \quad \text{or} \quad \sin \theta_c = \frac{n_2}{n_1} \tag{2.50}$$

Substitution of this result into Snell's law gives the angle of refraction to be 90°, and so a transmitted ray travels along the interface. If the angle of incidence is greater than θ_c (i.e. $\sin \theta_c > n_2/n_1$), then r_e will be complex, but $|r_e|$ will be unity, and total internal reflection still takes place. However, there will also be a transmitted wave. In order to study this in greater detail, let us consider the expression for the transmitted E field, reproduced here as (2.51):

$$\mathbf{E_t} = \mathbf{a}_z t_e E_0 \exp\left(j\beta_{2x}x + j\beta_{2y}y\right) \tag{2.51}$$

where β_{2y} is the phase constant in medium 2 resolved onto the y-axis. To evaluate E_t, we need to find β_{2x} and β_{2y} or, by implication, $\sin \theta_t$ and $\cos \theta_t$. If the incident ray hits the boundary at an angle greater than the critical angle, that is, $\theta_i > \theta_c$, then $\sin \theta_i > n_2/n_1$. If we substitute this into Snell's law, we find that $\sin \theta_t > 1$ which is physically impossible. We could work with hyperbolic functions at this point, but if we let $\sin \theta_t > 1$, then $\cos \theta_t$ will be imaginary, that is

$$\cos \theta_t = \sqrt{1 - \sin^2\theta_t} = jA \tag{2.52}$$

where A is a *real* number given by

$$A = \sqrt{\left(\frac{n_1}{n_2}\right)^2 \sin^2\theta_i - 1} \tag{2.53}$$

Thus the transmitted wave can be written as

$$\begin{aligned} \mathbf{E_t} &= \mathbf{a}_z t_e E_o \exp\left(j\beta_{2x}x + j^2 A\beta_2 y\right) \\ &= \mathbf{a}_z t_e E_o \exp\left(-A\beta_2 y\right) \exp\left(j\beta_{2x}x\right) \end{aligned} \tag{2.54}$$

This equation shows that an E field exists in the lower refractive index material *even though* total internal reflection takes place. As Eq. (2.54) shows, this field propagates without loss in the negative x-direction but undergoes attenuation as $\exp(-A\beta_2 y)$ along the y-axis, *at right angles to its direction of propagation.*

In order to find the transmitted power, we must also find the transmitted H field, previously given by Eq. (2.43):

$$\mathbf{H_t} = \left(-\mathbf{a}_x \cos \theta_t + \mathbf{a}_y \sin \theta_t\right) \frac{E_t}{Z_2}$$

where $E_t = t_e E_o \exp(-A\beta_2 y) \exp(j\beta_{2x}x)$. Substitution for $\cos \theta_t$ yields

$$\mathbf{H_t} = \left(-\mathbf{a}_x jA + \mathbf{a}_y \sin \theta_t\right) \frac{E_t}{Z_2} \tag{2.55}$$

As this equation shows, there are two components to the transmitted H field: a component along the x-axis that has a 90° phase shift, time-wise, with respect to the E field and a component along the y-axis. As the transmitted H field has two components, the transmitted power will also have two components.

As we have already seen in Sect. 2.1.2, the average power is given by

$$S_{av} = \frac{1}{2}\mathbf{E} \times \mathbf{H^*}$$

where $\mathbf{H^*} = \mathbf{H}\exp(-j[\omega t - \phi])$ and ϕ is the temporal phase angle between the individual E and H field components (90° for the x-axis component of H_t). So, with the fields given by (2.54) and (2.55), we have

$$S_{av} = -\mathbf{a}_x \frac{1}{2} \times \frac{E_t^2}{Z_2} \sin\theta_t - \mathbf{a}_y \frac{1}{2} \times \frac{AE_t^2}{Z_2} \exp\left(j\pi/2\right)$$
$$= -\mathbf{a}_x \frac{1}{2} \times \frac{E_t^2}{Z_2} \sin\theta_t - \mathbf{a}_y j \frac{1}{2} \times \frac{AE_t^2}{Z_2} \tag{2.56}$$

Thus it can be seen that the transmitted power has two components: an x-axis component (along the boundary) with the same properties as the transmitted E field and an imaginary component at right angles to the interface. This imaginary component is due to the 90° temporal phase shift between the E and H fields. (A similar relationship occurs between the voltage and current in reactive circuits.) The physical interpretation of this is that for the first quarter cycle, the E field is positive and the H field is negative, giving negative power flow. In the next quarter cycle, the E field is still positive, but the H field is also positive, and this gives positive power flow. Thus power flows to and from the boundary four times per complete cycle of the E or H field, and so no *net* power flows across the boundary along the y-axis.

The real part of the Poynting vector, from (2.56), is

$$S_{av} = -\mathbf{a}_x \frac{1}{2} \times \frac{E_t^2}{Z_2} \sin\theta_t \tag{2.57}$$

which shows that power flows *along* the boundary. So, although total internal reflection takes place, there is still a TEM wave propagating along the boundary – the *evanescent wave*. As we shall see in the following example, this wave is very tightly bound to the interface between the two media. We should note that the evanescent wave is why a cladding surrounds the core of an optical fibre. If air surrounds the core, total internal reflection will still take place, but the air cladding will contain the evanescent wave. Thus if we place another optical fibre close to the first fibre, the evanescent wave will cause coupling of power from one fibre to the other. This effect is desirable in optical couplers but is clearly undesirable when the fibres are tightly bound in a cable. By surrounding the core with a cladding of similar material, the power flow in the core is protected, both from coupling with adjacent fibres and from environmental effects.

Example
A TE wave, with a free-space wavelength of 600 nm, is propagating in a dielectric of refractive index 1.5. The wave hits a boundary with a second dielectric of refractive index 1.4, at an angle of 75° to a normal drawn perpendicular to the boundary. Determine the average transmitted power, and calculate the attenuation of the evanescent wave at a distance of one wavelength from the boundary.

Let us initially calculate the reflection coefficient, from which we can find the transmission coefficient. Now, r_e is given by Eq. (2.49):

(continued)

$$r_e = \frac{\cos\theta_i - \sqrt{\left(\frac{n_2}{n_1}\right)^2 - \sin^2\theta_i}}{\cos\theta_i + \sqrt{\left(\frac{n_2}{n_1}\right)^2 - \sin^2\theta_i}}$$

$$= \frac{0.26 - \sqrt{0.87 - 0.93}}{0.26 + \sqrt{0.87 - 0.93}}$$

$$= \frac{0.26 - j0.25}{0.26 + j0.25}$$

$$= -\frac{1}{2\phi}$$

where $\phi = \tan^{-1}(0.25/0.26) = 0.75$ rad. (This angle is the spatial phase change experienced by the E field on reflection.) Since $t_e = 1 + r_e$

$$t_e = \frac{0.26 + j0.25}{0.26 + j0.25} + \frac{0.26 - j0.25}{0.26 + j0.25}$$

$$= \frac{2 \times 0.26}{(0.26 + j0.25)}$$

$$= \frac{0.54}{-\phi}$$

Thus, $\mathbf{E_t}$ (Eq. 2.54) will be

$$\mathbf{E_t} = \mathbf{a}_z 0.54 E_o \exp\left(-3.93 \times 10^6 y\right) \exp\left(j1.57 \times 10^7 x\right)/-\phi$$

Hence the average transmitted power is

$$S_{av} = 5.4 \times 10^{-4} E_o{}^2 \exp\left(-7.86 \times 10^6 y\right)$$

We should note that, as the magnitude of r_e is unity, the reflected power is the same as the incident power. (The angle associated with r_e is simply the spatial phase shift experienced at the boundary.) Thus, we can say that the evanescent wave couples with, but takes no power from, the light travelling in the dielectric. It can, however, deliver power and most SM couplers rely on this property.

If we consider the average power at y equals one wavelength in the second dielectric, we find

(continued)

$$S_{av} = 5.4 \times 10^{-4} E_o{}^2 \exp(-3.38) \quad \text{for} \quad y = 430\,\text{nm}$$
$$= 18.4 \times 10^{-6} E_o{}^2$$

This power is 30 times less than that transmitted across the boundary – an attenuation of roughly 15 dB at a distance of one wavelength from the boundary. Clearly the evanescent wave is tightly bound to the interface between the two materials.

2.2.2 Propagation Modes: Ray-Path Analysis

In the previous section, we considered the reflection of a light ray at a single dielectric boundary. We showed that, provided the angle of incidence was greater than θ_c, total internal reflection would occur. It might be thought that any ray satisfying this requirement must propagate without loss. However, as we will shortly see, a light ray must satisfy certain conditions before it can successfully propagate. Figure 2.7 shows the situation we will analyse. In this diagram, the E field is drawn at right angles to the ray path. In order for the ray to propagate, the E field at A should be in phase with the E field at B, that is, *the ray must constructively interfere with itself.* If this is not the case, the fields will destructively interfere with each other, and the ray will simply die out. So, in order to maintain constructive interference at point B, the change in phase that the ray undergoes as it travels from A to B must be an integral number of cycles.

In going from A to B, the ray crosses the waveguide twice and is reflected off the boundary twice. We can find the phase change due to reflection off the boundary from the reflection coefficient. From (2.49), r_e is

$$r_e = \frac{\cos\theta_i - \sqrt{\left(\frac{n_2}{n_1}\right)^2 - \sin^2\theta_i}}{\cos\theta_i + \sqrt{\left(\frac{n_2}{n_1}\right)^2 - \sin^2\theta_i}}$$

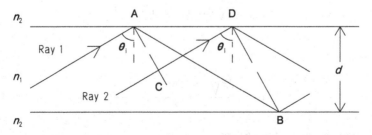

Fig. 2.7 Illustrative of the requirement for successful propagation of two TEM waves in a planar optical waveguide (CD $= a$, AB $= b$)

or, from the example at the end of the previous section

$$r_e = \frac{1}{-2\phi} \quad \text{where} \quad \phi = \tan^{-1} \frac{\sqrt{(n_1{}^2 \sin^2\theta_i - n_2{}^2)}}{n_1 \cos\theta_i} \tag{2.58}$$

Therefore, successful propagation occurs provided

$$2 \times 2d \times \beta_{1y} + 2 \times 2\phi = 2\pi N \tag{2.59}$$

where β_{1y} is the phase coefficient resolved onto the y-axis and N is a positive integer, known as the *mode number*. (Although we have taken upward travelling rays in Fig. 2.7, downward travelling rays will result in identical equations. These two rays make up a single *waveguide mode*.)

Now, if we substitute for ϕ (2.59) becomes

$$2d \times \beta_{1y} - 2\tan^{-1} \frac{\sqrt{(n_1{}^2 \sin^2\theta_i - n_2{}^2)}}{n_1 \cos\theta_i} = \pi N$$

or

$$\tan\left(\beta_{1y}d - \frac{\pi}{2}N\right) = \frac{\sqrt{(n_1{}^2 \sin^2\theta_i - n_2{}^2)}}{n_1 \cos\theta_i} \tag{2.60}$$

Now, $\beta_{1y} = \beta_1 \cos\theta_i$ and $\beta_1 = k_o n_1$ and so we can write (2.60) as

$$\tan\left(\beta_{1y}d - \frac{\pi}{2}N\right) = \frac{2\pi\sqrt{(n_1{}^2 \sin^2\theta_i - n_2{}^2)}}{\beta_{1y}\lambda_o} \tag{2.61}$$

From our previous discussions, the evanescent field undergoes attenuation as $\exp(-\alpha_2 y)$ and, from (2.53), we can write the attenuation factor as

$$\alpha_2 = \beta_2 A$$
$$= \beta_2\sqrt{(n_1/n_2)^2 \sin^2\theta_i - 1}$$

or

$$\alpha_2 = \frac{2\pi\sqrt{(n_1{}^2 \sin^2\theta_i - n_2{}^2)}}{\lambda_o} \tag{2.62}$$

Therefore we can write (2.61) as

$$\tan\left(\beta_{1y}d - \frac{\pi}{2}N\right) = \frac{\alpha_2}{\beta_{1y}} \tag{2.63}$$

Both (2.60) and (2.63) are known as *eigenvalue* equations. Solution of (2.63) will yield the values of β_{1y}, the *eigenvalues*, for which light rays will propagate, while solution of (2.60) will yield the permitted values of θ_i. Unfortunately, we can only solve these equations using graphical or numerical methods as the following example shows.

Example
Light of wavelength 1.3 μm is propagating in a planar waveguide of width 200 μm, depth 10 μm and refractive index 1.46, surrounded by material of refractive index 1.44. Find the permitted angles of incidence.

We can find the angle of incidence for each propagating mode by substituting these parameters into (2.60) and then solving the equation by graphical means. This is shown in Fig. 2.8, which is a plot of the left- and right-hand sides of (2.60), for varying angles of incidence, θ_i.

This graph shows that approximate values of θ_i are 88°, 86°, 84° and 82°, for mode numbers 0–3, respectively. Taking these values as a starting point, we can use numerical iteration to find the values of θ_i to any degree of accuracy. Thus the values of θ_i, to two decimal places, are 87.83°, 85.67°, 83.55° and 81.55°.

This analysis has shown that only those modes that satisfy the eigenvalue equation can propagate in the waveguide. We can estimate the number of modes by noting that the highest-order mode will propagate at the lowest angle of incidence. As this angle will have to be greater than or equal to the critical angle, we can use θ_c in (2.60) to give

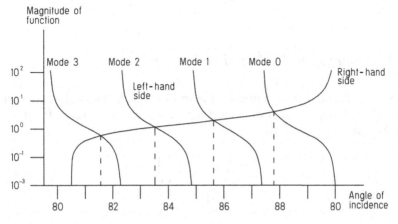

Fig. 2.8 Eigenvalue graphs for a planar dielectric waveguide

$$\tan \left(\frac{2\pi n_1}{\lambda_o} d \cos \theta_c - \frac{\pi}{2} N_{\max} \right) = 0$$

or

$$\frac{2\pi d \sqrt{n_1{}^2 - n_2{}^2}}{\lambda_o} = N_{\max} \frac{\pi}{2}$$

If we define a normalised frequency variable, V, as

$$V = \frac{2\pi d \sqrt{n_1{}^2 - n_2{}^2}}{\lambda_o} \qquad (2.64)$$

then the maximum number of modes will be

$$N_{\max} = \frac{2V}{\pi} = \frac{4d \sqrt{n_1{}^2 - n_2{}^2}}{\lambda_o} \qquad (2.65)$$

Equation (2.65) shows that the value of $N_{\max}$ is unlikely to be an integer, and so we must round it up to the nearest whole number. If we take the previous example, then $V = 5.82$, and so the number of propagating modes is 4. As can be seen from Fig. 2.8, there are only four solutions to the eigenvalue equation, so confirming the accuracy of (2.65). It should be noted that we can find the maximum propagating frequency, or wavelength, from V. We can also find the condition for single-mode operation from V. If $N_{\max} = 1$, V must be $\pi/2$, and we can find the waveguide depth from (2.65). In the next section we will apply Maxwell's equations to the planar dielectric waveguide. Although this will give us the same results as presented in this section, the treatment is more thorough, and it will help us when we come to consider propagation in optical fibre. (As the following analysis involves some complex mathematics, some readers may wish to neglect it on a first reading.)

2.2.3 Propagation Modes: Modal Analysis

To examine propagation in a planar dielectric waveguide thoroughly, we need to solve Maxwell's equations. If we assume that the dielectric is ideal, $\sigma = 0$ and we can write Maxwell's equations as

$$\nabla \times \mathbf{E} = -\mu \frac{\partial \mathbf{H}}{\partial t} \qquad (2.66a)$$

and

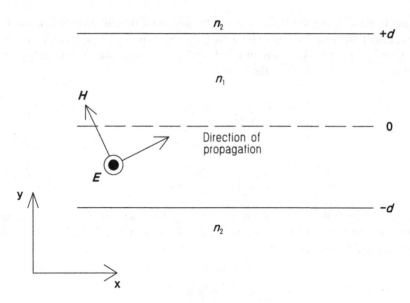

Fig. 2.9   A transverse electric wave propagating in a planar dielectric waveguide

$$\nabla \times \mathbf{H} = \varepsilon \frac{\partial \mathbf{E}}{\partial t} \tag{2.66b}$$

If we consider a transverse electric field propagating as in Fig. 2.9, we get

$$\nabla \times \mathbf{E} = \begin{vmatrix} \mathbf{a}_x & \mathbf{a}_y & \mathbf{a}_z \\ \dfrac{\partial}{\partial x} & \dfrac{\partial}{\partial y} & \dfrac{\partial}{\partial z} \\ 0 & 0 & E_z \end{vmatrix} = \frac{\partial}{\partial y} E_z \mathbf{a}_x - \frac{\partial}{\partial x} E_z \mathbf{a}_y = -\mu \left(\frac{\partial}{\partial t} H_x \mathbf{a}_x + \frac{\partial}{\partial t} H_y \mathbf{a}_y \right)$$

and

$$\nabla \times \mathbf{H} = \begin{vmatrix} \mathbf{a}_x & \mathbf{a}_y & \mathbf{a}_z \\ \dfrac{\partial}{\partial x} & \dfrac{\partial}{\partial y} & \dfrac{\partial}{\partial z} \\ H_x & H_y & 0 \end{vmatrix} = \left(\frac{\partial}{\partial x} H_y - \frac{\partial}{\partial y} H_x \right) \mathbf{a}_z = \varepsilon \frac{\partial}{\partial t} E_z \mathbf{a}_z$$

The fields are propagating in the x-direction and so we can write the x *component* of the E field as

$$\mathbf{E} = E \exp\left(j\omega t\right) \exp\left(-j\beta_x x\right) \mathbf{a}_z$$

with a similar expression for the H field. (Here β_x is the propagation coefficient of each individual mode along the x-axis.) Thus differentiation with respect to time results in multiplying the field strengths by $j\omega$, and differentiation with respect to x gives multiplication by $-j\beta_x$. (A similar argument applies to the y components of the E and H fields.) With this in mind, and by equating the vector coefficients, Maxwell's equations result in the following

$$\frac{\partial}{\partial y} E_z = -j\omega\mu H_x \tag{2.67a}$$

$$-j\beta_x E_z = j\omega\mu H_y \tag{2.67b}$$

$$-j\beta_x H_y - \frac{\partial}{\partial y} H_x = j\omega\varepsilon E_z \tag{2.67c}$$

Rather than derive an expression for the y component of the E field, we can rearrange (2.67a) and (2.67b) to give H_x and H_y in terms of E_z and substitute the results into (2.67c) to give

$$\frac{j\beta_x{}^2 E_z}{\omega\mu} + \frac{1}{j\omega\mu} \frac{\partial^2 E_z}{\partial y^2} = j\omega\varepsilon E_z$$

Hence we can write

$$\frac{\partial^2 E_z}{\partial y^2} = \left(\beta_x{}^2 - \omega^2\mu\varepsilon\right) E_z \tag{2.68}$$

which is valid in both the slab and the surrounding material.

We should note that the E field, as described by (2.68), is independent of time. So, the solution to (2.68) describes the *stationary distribution* of the *tangential E* field in the *vertical y-direction*. This distribution is independent of time because we are considering a waveguide, and reflections off the boundary walls result in a *standing wave* pattern that must satisfy (2.68).

We have three regions of interest: the surrounding material below $y = -d$; the surrounding material above $y = d$; and the slab material between $-d$ and $+d$. If we initially consider the solution to (2.68) in the surrounding material below $y = -d$, the E field must decrease exponentially away from the boundary (it is an evanescent wave). So, a possible solution is

$$E_z = K \exp\left(\alpha_2 y\right) \quad y < -d \tag{2.69a}$$

and, by using (2.67a),

$$H_x = \frac{-\alpha_2}{j\omega\mu} K \exp\left(\alpha_2 y\right) \quad y < -d \tag{2.69b}$$

where K is a constant and α_2 is the attenuation constant given by

$$\alpha_2{}^2 = \beta_x{}^2 - n_2{}^2 k_o{}^2 \tag{2.70}$$

(We can derive this equation by substituting Eq. 2.69a into Eq. 2.68.)

If we take $y > d$, we must have an exponential decrease in E_z with increasing y. So, another solution to (2.68) is

$$E_z = L \exp(-\alpha_2 y) \quad y > d \tag{2.71a}$$

and so

$$H_x = \frac{+\alpha_2}{j\omega\mu} L \exp(-\alpha_2 y) \quad y > d \tag{2.71b}$$

where L is another constant. If we consider the slab material, (2.68) becomes

$$\frac{\partial^2 E_z}{\partial y^2} = \left(\beta_{1x}{}^2 - n_1{}^2 k_o{}^2\right) E_z$$

with a possible solution given by

$$E_z = M \sin\left(\beta_{1y} + \varnothing\right) \quad -d < y < +d \tag{2.72a}$$

where M is a constant, β_{1y} is the phase coefficient in the slab material *resolved along the y-axis* and ϕ is a spatial phase shift that we have yet to determine. Thus, H_x will be given by

$$H_x = \frac{-M\beta_{1y}}{j\omega\mu} \cos\left(\beta_{1y} y + \varnothing\right) \quad -d < y < +d \tag{2.72b}$$

(This equation can be derived by substituting Eq. 2.72a into (2.68). We should also note that β_{1y}, β_{1x} and $n_1 k_o$ form a right-angled triangle. The proof of this is left as an exercise for the reader.)

So, we have six equations that describe the E_z and H_x fields in the slab and surrounding material. All of these equations have a number of constants that we can find using the boundary relations for the E and H fields. If we initially consider the lower interface, we have, from (2.69a) and (2.72a),

$$K \exp(-\alpha_2 d) = M \sin\left(-\beta_{1y} d + \phi\right) \quad \text{at} \ \ y = -d$$

for the continuity of the E_z field. After some simple manipulation, we get

$$K = M \sin\left(-\beta_{1y}d + \phi\right) \exp\left(\alpha_2 d\right)$$

The continuity of the H_x field gives, using (2.69b) and (2.72b)

$$\frac{-\alpha_2}{j\omega\mu} K \exp\left(-\alpha_2 d\right) = \frac{M\beta_{1y}}{j\omega\mu} \cos\left(-\beta_{1y}d + \varnothing\right)$$

and so

$$K = \frac{M\beta_{1y}}{\alpha_2} \cos\left(-\beta_{1y}d + \varnothing\right) \exp\left(\alpha_2 d\right)$$

If we equate the two equations for K, we get (after some rearranging)

$$\tan\left(-\beta_{1y}d + \varnothing\right) = \frac{\beta_{1y}}{\alpha_2}$$

or

$$-\beta_{1y}d + \varnothing = \tan^{-1}\left(\frac{\beta_{1y}}{\alpha_2}\right) + m'\pi \qquad (2.73)$$

where m' is an integer. Thus the phase angle ϕ is given by

$$\varnothing = \tan^{-1}\left(\frac{\beta_{1y}}{\alpha_2}\right) + \beta_{1y}d + m'\pi \qquad (2.74)$$

If we consider the upper interface, $y = +d$ we have, for the continuity of the E field,

$$L\exp\left(-\alpha_2 d\right) = M \sin\left(\beta_{1y}d + \varnothing\right)$$

After some rearranging we get

$$L = M \sin\left(\beta_{1y}d + \varnothing\right) \exp\left(\alpha_2 d\right)$$

The continuity of the H_x fields gives

$$\frac{\alpha_2}{j\omega\mu} L\exp\left(-\alpha_2 d\right) = -\frac{M\beta_{1y}}{j\omega\mu} \cos\left(\beta_{1y}d + \varnothing\right)$$

and so

$$L = -\frac{M\beta_{1y}}{\alpha_2} \cos\left(\beta_{1y}d + \varnothing\right) \exp\left(\alpha_2 d\right)$$

By equating the two values for L we get

$$M \sin\left(\beta_{1y}d + \varnothing\right) \exp\left(\alpha_2 d\right) = -\frac{M\beta_{1y}}{\alpha_2} \cos\left(\beta_{1y}d + \varnothing\right) \exp\left(\alpha_2 d\right)$$

Hence

$$\tan\left(\beta_{1y}d + \varnothing\right) = -\frac{\beta_{1y}}{\alpha_2}$$

and so another expression for ϕ is

$$\varnothing = -\tan^{-1}\left(\frac{\beta_{1y}}{\alpha_2}\right) - \beta_{1y}d + m''\pi \tag{2.75}$$

where m'' is another integer. If we equate (2.74) and (2.75), we get

$$\tan^{-1}\left(\frac{\beta_{1y}}{\alpha_2}\right) + \beta_{1y}d + m'\pi = -\tan^{-1}\left(\frac{\beta_{1y}}{\alpha_2}\right) - \beta_{1y}d + m''\pi$$

or

$$2\tan^{-1}\left(\frac{\beta_{1y}}{\alpha_2}\right) = -2\beta_{1y}d + m''\pi - m'\pi \tag{2.76}$$

We should note that m' and m'' are both integers that cover the range 0–∞ and so we can arbitrarily set m' to zero. Now

$$\tan^{-1}\left(\frac{\beta_{1y}}{\alpha_2}\right) = \frac{\pi}{2} - \tan^{-1}\left(\frac{\alpha_2}{\beta_{1y}}\right)$$

and so (2.76) becomes

$$\frac{\pi}{2} - \tan^{-1}\left(\frac{\alpha_2}{\beta_{1y}}\right) = -\beta_{1y}d + \frac{m''\pi}{2}$$

which can be written as

$$\tan\left(\beta_{1y}d - N\frac{\pi}{2}\right) = -\frac{\alpha_2}{\beta_{1y}} \qquad (2.77)$$

where N is an integer. The solution of this equation yields the values of β_{1y} for which light rays will propagate. Comparison with Eq. (2.63) shows that this method, based on Maxwell's equations, results in an equation that is identical to that derived using the ray model. (Although this derivation is more complicated than the ray-path model, it has introduced us to several important parameters as we shall see in the following sections.)

Let us now examine the cut-off condition for the planar waveguide. The maximum angle of incidence for any particular mode is the critical angle θ_c. If we can satisfy this condition, we find from our ray-path analysis that α_2 (Eq. 2.62) is zero. Thus the E_z component is not attenuated as it passes through the surrounding material, and the wave is not closely bound to the slab. If we take $\alpha_2 = 0$, we find from (2.70) that $\beta_x = \beta_2$ and so the phase coefficient of the propagating mode is identical to that of the surrounding material. This is known as the *cut-off condition*, and it represents the minimum value of β for which any mode will propagate. If the waveguide is operating well away from the cut-off condition, the propagating modes will be tightly bound to the slab, and so we can intuitively reason that $\beta_x = \beta_1$. Thus we can see that each propagating mode must have $\beta_2 < \beta_x < \beta_1$. We can define a binding parameter, b, as

$$b = \frac{\beta_x^{\,2} - n_2^{\,2}k_o^{\,2}}{n_1^{\,2}k_o^{\,2} - n_2^{\,2}k_o^{\,2}} \qquad (2.78)$$

and so, with $n_2 k_o < \beta_x < n_1 k_o$ the range of b will be 0 for loosely bound modes, to 1 for tightly bound modes (Fig. 2.9).

(An alternative way of viewing this binding parameter is to let each mode have an *effective refractive index*, n_{eff}. Thus (2.78) will become

$$b = \frac{n_{eff}^{\,2} - n_2^{\,2}}{n_1^{\,2} - n_2^{\,2}}$$

Re-casting (2.78) in this form can be useful in that it shows that the velocity of the propagating mode is bounded by the velocity in the core, for tightly bound modes, and the velocity in the cladding material, for loosely bound modes. We will use the term effective refractive index very shortly.)

We can find the cut-off wavelength of any guide by substituting $\alpha_2 = 0$ into (2.77) to give

$$\tan\left(\beta_{1y}d - N\frac{\pi}{2}\right) = 0$$

which implies

$$\beta_{1y}d = N\frac{\pi}{2} \tag{2.79}$$

As $\beta_{1y} = \beta_1\cos\theta_c$, $\cos^2\theta_c = 1 - \sin^2\theta_c$ and $\sin\theta_c = n_2/n_1$, we can write

$$\beta_1 d \sqrt{\left(1 - \frac{n_2{}^2}{n_1{}^2}\right)} = N\frac{\pi}{2}$$

or

$$\frac{2\pi n_1 d}{\lambda_{co}} \sqrt{\left(\frac{n_1{}^2 - n_2{}^2}{n_1{}^2}\right)} = N\frac{\pi}{2}$$

Thus the cut-off wavelength for a particular mode is given by

$$\lambda_{co} = \frac{4d}{N}\sqrt{(n_1{}^2 - n_2{}^2)} \tag{2.80}$$

If we consider the lowest-order mode $(N = 0)$, we have a λ_{co} of infinity, and so there is, theoretically, no cut-off wavelength for the lowest-order mode. If we take $N = 1$, we can determine the waveguide depth that just results in the first-order mode being cut off.

Example
Light of wavelength 1.3 μm is propagating in a planar waveguide of width 200 μm and refractive index 1.46, surrounded by material of refractive index 1.44. Determine the waveguide depth for single-mode operation.
   Let us consider the $N = 1$ mode and allow this mode to be just cut off. The cut-off condition for this mode is

$$1.3 \times 10^{-6} = 4d\sqrt{\left(1.46^2 - 1.44^2\right)}$$

and so the maximum waveguide depth is 2.7 μm.

Before we finish this particular section, let us return to the cut-off condition for any particular mode (2.79):

$$\beta_{1y}d = N\frac{\pi}{2}$$

This can be written as

$$V = N\frac{\pi}{2} \tag{2.81}$$

where

$$V = \frac{2\pi d}{\lambda_o}\sqrt{(n_1{}^2 - n_2{}^2)} \tag{2.82}$$

is known as the *V value* of the waveguide. Equations (2.81) and (2.82) are identical to those we derived using the ray-path analysis (Eqs. 2.64 and 2.65). We can also express V as

$$V^2 = (\alpha_2 d)^2 + (\beta_{1y} d)^2 \tag{2.83}$$

This can be easily proved by noting that

$$\alpha_2{}^2 = \beta_x{}^2 - n_2{}^2 k_o{}^2$$

and

$$\beta_{1y}{}^2 = n_1{}^2 k_o{}^2 - \beta_x{}^2$$

Thus

$$V^2 = d^2\left(\beta^2 - \beta_2{}^2 + \beta_1{}^2 - \beta^2\right)$$
$$= d^2\left(n_1{}^2 k_o{}^2 - n_2{}^2 k_o{}^2\right)$$

and so

$V = \frac{2\pi d\sqrt{n_1{}^2 - n_2{}^2}}{\lambda_o}$ which is identical to (2.64).

In this section we have applied Maxwell's equations to a planar dielectric waveguide. We finished by showing that we can reduce the number of propagating modes by decreasing the waveguide thickness. In particular, if the V value of the waveguide is less than $\pi/2$, then only the zero-order mode can propagate (so-called monomode or *single-mode*, SM, operation). As we shall see in the next section, single-mode operation helps to reduce the total dispersion.

2.2.4 Modal Dispersion: Ray-Path Analysis

We have already seen that only a certain number of modes can propagate. Each of these modes carries the modulation signal and, as each one is incident on the boundary at a different angle, they will each have their own individual propagation times. In a digital system, the net effect is to smear out the pulses, and so this is a form of dispersion – *modal dispersion.*

The difference in arrival time, δt, between the fastest and slowest modes will be given by

$$\delta t = t_{\max} - t_{\min} \tag{2.84}$$

where $t_{\max}$ and $t_{\min}$ are the propagation times of the highest- and lowest-order modes, respectively. As it is the modulation envelope that carries the information, we can find $t_{\max}$ and $t_{\min}$ by dividing the waveguide length by the *axial* components of the group velocities. As this requires knowledge of θ_i for the various waveguide modes, it may not be a practical way of estimating δt.

We can obtain an indication of the dispersion by approximating the angle of incidence for the highest-order mode to θ_c and that of the zero-order mode to $90°$. Thus

$$t_{\min} = \frac{LN_{g1}}{c} \quad \text{and} \quad t_{\max} = \frac{LN_{g1}}{c \sin \theta_c} = \frac{LN_{g1}{}^2}{cN_{g2}}$$

where L is the length of the waveguide, and we have used the group refractive indices N_{g1} and N_{g2}. Therefore δt will be

$$\delta t = \frac{LN_{g1}}{cN_{g2}} \left(N_{g1} - N_{g2} \right) \tag{2.85}$$

Now, if we take $N_{g1}/n_1 \approx N_{g2}/n_2$, then the dispersion *per unit length* will be

$$\frac{\delta t}{L} = \frac{N_{g1}}{cN_{g2}} \left(N_{g1} - N_{g2} \right) = \frac{N_{g1}}{cn_2} (n_1 - n_2) = \frac{N_{g1}\delta n}{c}$$

or

$$\sigma_{\text{mod}} = \frac{N_{g1}\delta n}{c} \tag{2.86}$$

where δn is the *fractional refractive index difference* and σ_{mod} is the dispersion per unit length. We should note that each individual mode will also suffer from material dispersion. Thus, when we resolve the transit time of each mode on to the fibre axis, we should also take into account the material dispersion. Fortunately, in most multimode waveguides, the material dispersion is far less than the modal dispersion, and so we can generally ignore its effects.

Example
**Light of wavelength 850 nm is propagating in a waveguide of 10 μm
depth, with refractive indices $n_1 = 1.5$ and $n_2 = 1.4$, and group refractive
indices $N_{g1} = 1.64$ and $N_{g2} = 1.53$. Determine the modal dispersion.**
  If we use Eq. (2.86), we find

$$\sigma_{mod} = \frac{1.64}{3 \times 10^8} \times \frac{1.5 - 1.4}{1.4} = 0.39\,ns/km$$

  If we use modal analysis, we find that there are thirteen modes, and the
angles of incidence for the zero- and thirteenth-order modes are 89.21° and
69.96°, respectively. This method results in $\sigma_{mod} = 0.35$ ns/km, and so we can
see that the error in using (2.86) is small. If the waveguide is single-mode, then
σ_{mod} reduces to zero.

2.2.5 Modal Dispersion: Modal Analysis

The analysis just presented used the ray-path approximation to determine the modal
dispersion. Although the error in using this approximation is very small, a more
rigorous treatment using mode theory will aid us when we consider optical fibre. To
help us in our analysis, we will use the binding parameter, b, previously defined by
(2.78):

$$b = \frac{\beta^2 - n_2^2 k_o^2}{n_1^2 k_o^2 - n_2^2 k_o^2} \tag{2.87}$$

We can rearrange (2.87) to give

$$\begin{aligned}\beta &= \sqrt{\left[n_2^2 k_o^2 + b\left(n_1^2 k_o^2 - n_2^2 k_o^2\right)\right]} \\ &= \frac{2\pi n_2}{\lambda_o} \sqrt{\left[1 + b\left(\frac{n_1^2 - n_2^2}{n_2^2}\right)\right]}\end{aligned} \tag{2.88}$$

  Most optical waveguides are fabricated with $n_1 = n_2$ (a condition known as
weakly guiding [4]) and so (2.88) becomes

$$\begin{aligned}\beta &= \frac{2\pi n_2}{\lambda_o} \sqrt{(1 + 2b\delta n)} \\ &= \frac{\omega n_2}{c} \sqrt{(1 + 2b\delta n)} \\ &= \frac{\omega n_2}{c} (1 + b\delta n)\end{aligned} \tag{2.89}$$

where we have used the binomial expansion. Now, modes travel at the group velocity, v_g, and so the time taken for a mode to travel a unit length, τ, is given by

$$\tau = \frac{1}{v_g} = \frac{d\beta}{d\omega} = (1 + b\delta n)\frac{d}{d\omega}\left[\frac{\omega n_2}{c}\right] + \left[\frac{\omega n_2}{c}\right]\frac{d}{d\omega}(1 + b\delta n) \qquad (2.90)$$

The second term in (2.90) is small enough to be neglected when compared to the first term and so

$$\tau = \frac{1}{c} = (1 + b\delta n)\left[n_2 + \omega\frac{dn_2}{d\omega}\right] = \frac{N_{g2}}{c}(1 + b\delta n) \qquad (2.91)$$

We have already seen that the propagation constant for a particular mode has a range given by $\beta_2 < \beta < \beta_1$ and so b must lie in the range $0 < b < 1$. Thus, the difference in propagation time between the highest- and lowest-order modes is

$$\begin{aligned}
\frac{\delta t}{L} &= \frac{N_{g2}}{c}(1 + b_1\delta n - 1 - b_0\delta n) \\
&= \frac{N_{g2}\delta n}{c}(b_1 - b_0) \\
&= \frac{N_{g2}\delta n}{c}
\end{aligned} \qquad (2.92)$$

where we have taken $b_1 = 1$, and $b_0 = 0$. This should be compared to Eq. (2.86) obtained using ray-path analysis. One obvious difference is that (2.86) uses the group refractive index in the slab material, whereas (2.92) uses the group refractive index in the surrounding material. However, we should remember that this analysis uses the weakly guiding approximation, and so $N_{g1} = N_{g2}$ and (2.92) and (2.86) become equivalent.

2.2.6   Waveguide Dispersion: Ray-Path and Modal Analysis

As well as suffering from modal and material dispersion, a propagating signal will also undergo *waveguide dispersion*. As we will see, waveguide dispersion results from the variation of propagation coefficient, and hence allowed angle of incidence, with wavelength. (When we considered propagation in Sect. 2.2.2, we found that only those modes that satisfy the eigenvalue Eq. (2.60) will propagate successfully. As the refractive index of the waveguide material is present in (2.60), θ_i will vary if n changes with wavelength.) In common with the previous sections, we will initially use a ray-path analysis. However, this will limit us somewhat, and so we must resort to a modal analysis. With this in mind, some readers can neglect the latter part of this section on a first reading.

By following an analysis similar to that used in Sect. 2.1.3, the *axial* transit time per unit length per unit of source line width, is

$$
\begin{aligned}
\tau &= \frac{d\beta_{1x}}{d\omega} = -\frac{\lambda_o{}^2}{2\pi c} \times \frac{d}{d\lambda}\beta_{1x} \\
&= -\frac{\lambda_o{}^2}{2\pi c} \times \frac{d}{d\lambda}\left[\frac{2\pi n_1 \sin\theta_i}{\lambda_o}\right] \\
&= \frac{\sin\theta_i}{c}\left[n_1 - \lambda_o\frac{dn_1}{d\lambda}\right] - \frac{n_1\lambda_o}{c} \times \frac{d}{d\lambda}\sin\theta_i
\end{aligned}
\tag{2.93}
$$

The first term in (2.93) leads to the material dispersion resolved on to the *x*-axis (refer to Eq. 2.29). We should expect this because each individual mode will suffer from material dispersion. However, the second term in (2.93) leads to the waveguide dispersion, σ_{wg}. This is due to the waveguide propagation constants, and hence the permitted angles of incidence, varying with wavelength.

Let us now turn our attention to a modal analysis. In the previous section we expressed the propagation coefficient, of a particular mode, as (Eq. 2.89)

$$
\beta = \frac{\omega n_2}{c}(1 + b\delta n)
$$

or

$$
\begin{aligned}
\beta &= n_2 k_o(1 + b\delta n) \\
&= \beta_2(1 + b\delta n)
\end{aligned}
$$

Now, material dispersion results from the variation of *group refractive index* with wavelength. We have already defined k_o as the free-space propagation coefficient, and so the group refractive index of a particular mode can be written as

$$
\begin{aligned}
N_g &= \frac{d\beta}{dk_o} = \frac{d\beta_2}{dk_o} + \frac{d}{dk_o}(\beta_2 b\delta n) \\
&= \frac{d\beta_2}{dk_o} + \frac{d}{d\beta_2}(\beta_2 b\delta n)\frac{d\beta_2}{dk_o}
\end{aligned}
\tag{2.94}
$$

The *V* value of the guide is given by Eq. (2.82)

$$
V = \frac{2\pi d}{\lambda_o}\sqrt{(n_1{}^2 - n_2{}^2)} \approx \beta_2 d\sqrt{2\delta n}
\tag{2.95}
$$

where we have used the weakly guiding approximation ($n_1 \approx n_2$) and the binomial expansion. If we take the variation of δn with β_2 to be small, we can use (2.95) in (2.94) to give

$$N_g = \frac{d\beta_2}{dk_o} + \frac{d}{dV}(Vb\delta n)\frac{d\beta_2}{dk_o}$$

$$= \frac{d\beta_2}{dk_o}\left[1 + \frac{d}{dV}(Vb\delta n)\right]$$

$$= \frac{d\beta_2}{dk_o}\left[1 + \delta n\frac{d}{dV}(Vb)\right]$$

where we have neglected the variation of δn with V. The term outside the bracket is the group refractive index of the surrounding material, and so we can write the group refractive index of a propagating mode as

$$N_g = N_{g2}\left[1 + \delta n\frac{d}{dV}(Vb)\right] \tag{2.96}$$

Dispersion results from the variation of this group refractive index with wavelength. We can see from (2.96) that the dispersion will depend on two components: the first yields the material dispersion; however, the second term involves the mode-dependent term $d(Vb)/dV$. This is the *waveguide dispersion*. We should note that even if the guide is single-mode, waveguide dispersion will still be present. This is because both V and b are wavelength dependent, and so the linewidth of the optical source will contribute to waveguide and material dispersion.

We have now completed our study of dispersion in planar optical waveguides. We have found that optical signals are distorted by three mechanisms: modal dispersion caused by the dimensions of the waveguide allowing many modes to propagate; material dispersion caused by the group refractive index of the waveguide varying with wavelength; and waveguide dispersion caused by the waveguide propagation parameters being dependent on wavelength. It is worth remembering at this point that planar optical waveguides are usually very short in length, and so the dispersion effects we have been studying are not normally significant. However, these studies have introduced us to some very important concepts that will help our investigation of propagation in optical fibre.

2.2.7 Numerical Aperture

Figure 2.10 shows two light rays entering a planar waveguide. Refraction of both rays occurs on entry; however, ray 1 fails to propagate in the guide because it hits the boundary at an angle less than θ_c. On the other hand, ray 2 enters the waveguide at an angle θ_i and then hits the boundary at θ_c; thus it will propagate successfully. If θ_i is the maximum angle of incidence, then the *numerical aperture*, *NA*, of the waveguide is equal to the sine of θ_i.

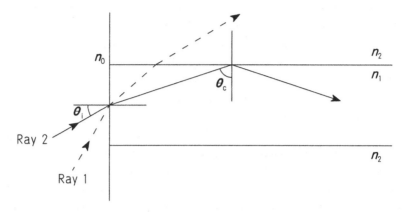

Fig. 2.10 Construction for the determination of the numerical aperture

We can find the NA by applying Snell's law to ray 2. Thus,

$$n_o \sin \theta_i = n_1 \sin \left(\frac{\pi}{2} - \theta_c\right) = n_1 \cos \theta_c$$

$$= n_1 \sqrt{1 - \frac{n_2{}^2}{n_1{}^2}}$$

Therefore

$$\sin \theta_i = \frac{1}{n_o} \sqrt{n_1{}^2 - n_2{}^2} \tag{2.97}$$

If the guide is in air, then

$$\mathrm{NA} = \sin \theta_i = \sqrt{n_1{}^2 - n_2{}^2} \tag{2.98}$$

A large NA results in efficient coupling of light into the waveguide. However, a high NA implies that $n_1 \gg n_2$ which results in a large amount of modal dispersion, so limiting the available bandwidth.

2.3 Propagation in Optical Fibres

So far we have only considered propagation in an infinite dielectric block and a planar dielectric waveguide. In this section we shall consider a cylindrical waveguide – the optical fibre. Light rays propagating in the fibre core fall into one

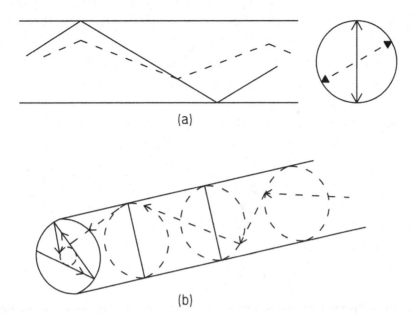

(a)

(b)

Fig. 2.11 Propagation of (**a**) meridional and (**b**) skew rays in the core of a step-index, MM, optical fibre

of two groups. The first group consists of those light rays which pass through the axis of the core. Such rays are known as *meridional rays*, and Fig. 2.11a shows the passage of two of these rays propagating in a step-index fibre. With a little thought, it should be apparent that we can regard meridional rays as equivalent to the rays we considered in the planar dielectric.

The second group consists of those rays that never pass through the axis, known as *skew rays*. As Fig. 2.11b shows, these rays do not fully utilise the area of the core. As skew rays travel significantly farther than meridional rays, they generally undergo higher attenuation.

In the following section, we will apply Maxwell's equations to a cylindrical waveguide. Unfortunately, this will involve us in some rather complex mathematics which some readers can neglect on a first reading. However, the full solution does yield some very important results, and these are quoted in later sections. In common with the planar waveguide, we will find the condition for single-mode operation. We will then go on to study the dispersion characteristics of a cylindrical waveguide.

2.3.1 Propagation in Step-Index Optical Fibres

When we considered the planar waveguide, we solved Maxwell's equations using Cartesian coordinates. Now that we are considering a cylindrical waveguide, we must use *cylindrical coordinates*, as shown in Fig. 2.12.

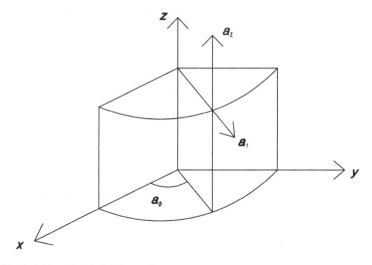

Fig. 2.12 Definition of cylindrical coordinate set

In the following derivations, mention is made of *waveguide modes*. As we saw when we considered the planar waveguide, only certain modes (ray paths) could propagate successfully. A similar situation exists in the cylindrical waveguide, as we will see when we come to solve the relevant eigenvalue equation.

Maxwell's equations, as applied to an ideal insulator such as glass, can be written as

$$\nabla \times \mathbf{E} = -\mu \frac{\partial \mathbf{H}}{\partial t} \tag{2.99a}$$

and

$$\nabla \times \mathbf{H} = \varepsilon \frac{\partial \mathbf{E}}{\partial t} \tag{2.99b}$$

If we assume that the $\mathbf{E}$ and $\mathbf{H}$ fields propagate in the positive z-direction with negligible attenuation, we can write

$$\mathbf{E} = E_0(r, \varnothing) e^{j(\omega t - \beta z)} \tag{2.100a}$$

$$\mathbf{H} = H_0(r, \varnothing) e^{j(\omega t - \beta z)} \tag{2.100b}$$

where we have used the phasor representation of $\mathbf{E}$ and $\mathbf{H}$ and β is the phase constant of any particular mode. If we substitute (2.100a) into (2.99a), we get

$$\nabla \times \mathbf{E} = \begin{vmatrix} \mathbf{a_r}1/r & \mathbf{a_\varnothing} & \mathbf{a_z}1/r \\ \dfrac{\partial}{\partial r} & \dfrac{\partial}{\partial \varnothing} & \dfrac{\partial}{\partial z} \\ E_r & rE_\varnothing & E_z \end{vmatrix} = -\mu\left(\frac{\partial}{\partial t}H_r\mathbf{a_r} + \frac{\partial}{\partial t}H_\varnothing\mathbf{a_\varnothing} + \frac{\partial}{\partial t}H_z\mathbf{a_z}\right)$$

By equating unit vectors, and performing the differentiation with respect to time, we get

$$\frac{1}{r}\left(\frac{\partial}{\partial \varnothing}E_z + j\beta rE_\varnothing\right) = -j\omega\mu H_r \tag{2.101a}$$

$$-\left(\frac{\partial}{\partial r}E_z + j\beta E_r\right) = -j\omega\mu H_\varnothing \tag{2.101b}$$

and

$$\frac{1}{r}\left(\frac{\partial}{\partial r}rE_\varnothing - \frac{\partial}{\partial \varnothing}E_r\right) = -j\omega\mu H_z \tag{2.101c}$$

By following a similar procedure with the H field, we get

$$\frac{1}{r}\left(\frac{\partial}{\partial \varnothing}H_z + j\beta rH_\varnothing\right) = j\omega\varepsilon E_r \tag{2.102a}$$

$$-\left(\frac{\partial}{\partial r}H_z + j\beta H_r\right) = j\omega\varepsilon E_\varnothing \tag{2.102b}$$

and

$$\frac{1}{r}\left(\frac{\partial}{\partial r}rH_\varnothing - \frac{\partial}{\partial \varnothing}H_r\right) = j\omega\varepsilon E_z \tag{2.102c}$$

Now, if we rearrange (2.102b) to give H_r in terms of E_ϕ and H_z and substitute the result into (2.101a), we can write

$$E_\varnothing = -\frac{j}{\omega^2\mu\varepsilon - \beta^2}\left(\frac{\beta}{r}\frac{\partial}{\partial \phi}E_z - \mu\frac{\partial}{\partial r}H_z\right) \tag{2.103a}$$

Similarly

$$E_r = -\frac{j}{\omega^2\mu\varepsilon - \beta^2}\left(\beta\frac{\partial}{\partial r}E_z + \frac{\omega\mu}{r}\frac{\partial}{\partial \varnothing}H_z\right) \tag{2.103b}$$

$$H_\emptyset = -\frac{j}{\omega^2\mu\varepsilon - \beta^2}\left(\frac{\beta}{r}\frac{\partial}{\partial\phi}H_z - \omega\varepsilon\frac{\partial}{\partial r}E_z\right) \qquad (2.103c)$$

and

$$H_r = -\frac{j}{\omega^2\mu\varepsilon - \beta^2}\left(\beta\frac{\partial}{\partial r}H_z - \frac{\omega\varepsilon}{r}\frac{\partial}{\partial\emptyset}E_z\right) \qquad (2.103d)$$

We can now substitute (2.103d) and (2.103c) into (2.102c) to give

$$\frac{\partial^2}{\partial r^2}E_z + \frac{1}{r}\frac{\partial}{\partial r}E_z + \frac{1}{r^2}\frac{\partial^2}{\partial\emptyset^2}E_z + \left(\omega^2\mu\varepsilon - \beta^2\right)E_z = 0 \qquad (2.104)$$

and, by substituting (2.103a) and (2.103b) into (2.102c), we get

$$\frac{\partial^2}{\partial r^2}H_z + \frac{1}{r}\frac{\partial}{\partial r}H_z + \frac{1}{r^2}\frac{\partial^2}{\partial\emptyset^2}H_z + \left(\omega^2\mu\varepsilon - \beta^2\right)H_z = 0 \qquad (2.105)$$

Equations (2.104) and (2.105) are the wave equations for the E and H fields as applied to a circular waveguide. We should note that these equations apply equally to the core and cladding of step-index optical fibre. As E and H are travelling waves, a general solution to these equations is

$$E_z(t, r, \emptyset, z) = Ax(r)y(\emptyset)e^{j(\omega t - \beta z)} \qquad (2.106)$$

and

$$H_z(t, r, \emptyset, z) = Bx(r)y(\emptyset)e^{j(\omega t - \beta z)} \qquad (2.107)$$

where A and B are constants and $x(r)$ and $y(\phi,)$ are yet to be determined.

To find $y(\phi,)$, let us consider the variation of E_z, with ϕ. A complete rotation of E_z occurs as ϕ goes from 0 to 2π. Hence E_z must be the same at all multiples of 2π, that is, E_z *must be periodic with* ϕ. Thus we can say

$$y(\emptyset) = e^{jv\emptyset}$$

where v is an integer. We can now substitute for $y(\phi)$ into (2.104) and use the result in the wave equation for E, to give (after some cancellation)

$$\frac{\partial^2}{\partial r^2}x(r) + \frac{1}{r}\frac{\partial}{\partial r}x(r) + \left[\left(\omega^2\mu\varepsilon - \beta^2\right) - \frac{v^2}{r^2}\right]x(r) = 0 \qquad (2.108)$$

This equation is known as Bessel's differential equation, and the solution uses Bessel functions with the bracketed term in (2.108) as the argument.

To solve (2.108) we must consider two regions of interest: $r < a$ (the core of the fibre) and $r > a$ (the cladding of the fibre) where a is the fibre core radius. If we first consider the core and let $a \to 0$, there must be finite solutions to (2.108). (After all, light propagates in the core, and so there must be solutions to (2.108).) The functions that satisfy this condition are *Bessel functions of the first kind*. The solution to (2.108) then becomes

$$x(r) = J_v(ur)$$

and so

$$E_z(t, r, \varnothing, z) = A J_z(ur)e^{jv\varnothing}e^{j(\omega t - \beta z)} \tag{2.109}$$

and

$$H_z(t, r, \varnothing, z) = B J_z(ur)e^{jv\varnothing}e^{j(\omega t - \beta z)} \tag{2.110}$$

where $u^2 = \omega^2 \mu \varepsilon - \beta^2$ and v is the order of the Bessel function. By noting that $1/\sqrt{\mu\varepsilon} = n/c$ we can write

$$u^2 = \left[\frac{2\pi n_1}{\lambda_o}\right]^2 - \beta^2$$

or

$$u^2 = \beta_1^2 - \beta^2$$

Let us now consider the cladding modes. Outside the core the E and H fields must decay away to zero for large radius. (This is a necessary condition for evanescent waves which we first encountered in Sect. 2.2.1.) Now, if we let the radius tend to infinity, the Bessel functions of the first kind are finite and *not zero*. Hence we must use the *modified Bessel functions of the second kind* which give zero field at large radius. Thus

$$x(r) = K_v(wr)$$

and so

$$E_z(t, r, \phi, z) = C K_v(wr)e^{jv\varnothing}e^{j(\omega t - \beta z)} \tag{2.111}$$

and

$$H_z(t, r, \phi, z) = D K_v(wr) e^{jv\varnothing} e^{j(\omega t - \beta z)} \tag{2.112}$$

where the argument of the function is wr. We can find an expression for w by noting that as $a \to \infty$, $K_v(wa) \to e^{-wa}$. To ensure that $K_v(wa)$ tends to zero, we must have $w > 0$ and so

$$w^2 = \beta^2 - \left[\frac{2\pi n_2}{\lambda_o}\right]^2$$

or

$$w^2 = \beta^2 - \beta_2{}^2$$

We can immediately see that w^2 is different from u^2 in that the order of the subtraction is reversed, that is

$$w^2 = -\left(\beta_2{}^2 - \beta^2\right) \quad \text{whereas} \quad u^2 = \left(\beta_1{}^2 - \beta^2\right)$$

(This is of importance when we use the boundary conditions for $E_{\phi 2}$ and $H_{\phi 2}$.) We can also perform a simple check on the values of u^2 and w^2 by noting

$$\beta_2{}^2 < \beta^2 < \beta_1{}^2 \quad \text{provided} \quad n_2 < n_1$$

Thus we can see that $J_v(ur)$ and $K_v(wr)$ satisfy the conditions placed on the E and H fields in the core and cladding. In passing, it is interesting to note that the lowest-order mode will have a phase coefficient of β_1, that is, it will be totally confined to the core, whereas the highest-order mode will propagate with β_2, that is, it will travel in the cladding – an evanescent wave. This is an identical situation to that encountered when we considered the planar optical waveguide.

The constants A, B, C and D can be found by applying boundary conditions to the E and H fields, that is, the tangential components of the $E_{z,\phi}$ and $H_{z,\phi}$ fields are continuous across the boundary between the core and the cladding. So, for the tangential E, field, we have

$$E_{z1} = E_{z2}$$

or

$$E_{z1} - E_{z2} = 0$$

Hence

$$A J_v(ua) - C K_v(wa) = 0 \tag{2.113}$$

By a similar process, the continuity of the H field yields

$$BJ_v(ua) - DK_v(wa) = 0 \tag{2.114}$$

As regards the E component, we can substitute (2.109) and (2.110) into (2.113) to give E, for $r < a$, and substitute (2.111) and (2.112) into (2.113) to give E, for $r > a$. Thus, the boundary condition $E_{\phi 1} - E_{\phi 2} = 0$ becomes

$$-\frac{j}{u^2} \left[A\frac{jv\beta}{a} J_v(ua) - B\omega\mu u J_v'(ua) \right]$$
$$+\frac{j}{w^2} \left[C\frac{jv\beta}{a} K_v(wa) - D\omega\mu w K'(wa) \right] = 0 \tag{2.115}$$

where the prime symbols – ' – denotes differentiation with respect to radius. (The apparent reversal in the sign of $E_{\phi 2}$ results from the difference between u^2 and w^2 noted earlier.) By applying the same procedure to the H_ϕ field, we can write

$$-\frac{j}{u^2} \left[B\frac{jv\beta}{a} J_v(ua) + A\omega\varepsilon_1 u J_v'(ua) \right]$$
$$-\frac{j}{w^2} \left[D\frac{jv\beta}{a} K_v(wa) + C\omega\varepsilon_2 w K'(wa) \right] = 0 \tag{2.116}$$

So, we have four Eqs. (2.113), (2.114), (2.115) and (2.116) and four unknowns A, B, C and D. If we express these equations in the form of a matrix, the determinant will yield the permitted values of the propagation constant β. Thus

$$\begin{vmatrix} J_v(ua) & 0 & -K_v(wa) & 0 \\ \dfrac{\beta v}{au^2} J_v(ua) & \dfrac{j\omega\mu}{u} J_v'(ua) & \dfrac{\beta v}{aw^2} K_v(wa) & \dfrac{j\omega\mu}{w} K_v'(wa) \\ 0 & J_v(ua) & 0 & -K_v(wa) \\ \dfrac{-j\omega\varepsilon_1}{u} J_v'(ua) & \dfrac{\beta v}{au^2} J_v(ua) & \dfrac{-j\omega\varepsilon_2}{w} K_v'(wa) & \dfrac{\beta v}{aw^2} K_v(wa) \end{vmatrix} = 0$$

This rather complex determinant results in the following eigenvalue equation:

$$\left[\frac{J_v'(ua)}{uJ_v(ua)} + \frac{K_v'(wa)}{wK_v(wa)} \right] \left[\beta_1{}^2 \frac{J_v'(ua)}{uJ_v(ua)} + \beta_2{}^2 \frac{K_v'(wa)}{wK_v(wa)} \right] =$$
$$\left[\frac{\beta v}{a} \right]^2 \left[\frac{1}{u^2} + \frac{1}{w^2} \right]^2 \tag{2.117}$$

The complete solution to this equation will yield the values of β for which a mode will propagate in the fibre. Unfortunately, (2.117) can only be solved by graphical/numerical techniques similar to those used when we considered the planar waveguide. (We will leave the solution to this equation until we consider a particular fibre type in Sect. 2.3.3.)

Before we go on to examine dispersion in optical fibre, let us return to the four equations that link A, B, C and D – Eqs. (2.113), (2.114), (2.115) and (2.116). From (2.113) and (2.114) we can write

$$\frac{A}{C} = \frac{K_v(wa)}{J_v(ua)} = \frac{B}{D}$$

which simply relate the E and H fields inside the core to those in the cladding. However, (2.115) and (2.116) result in

$$\frac{A}{B} = \frac{-j\omega\mu}{\beta v}\left[\frac{J_v'(ua)}{uaJ_v(ua)} + \frac{K_v'(wa)}{waK_v(wa)}\right]\left[\frac{u^2w^2a^2}{u^2+w^2}\right]$$

which indicates that the E and H fields inside the core are linked. This is an important result because it shows that, unlike rectangular waveguides, circular waveguides can support *hybrid modes* as well as the more familiar transverse electric and transverse magnetic modes. These hybrid modes are designated as *EH* or *HE* depending on the relative magnitude of the E and H field components transverse to the fibre axis. (We can visualise these hybrid modes as being the skew rays shown in Fig. 2.11.)

2.3.2 Dispersion in Cylindrical Waveguides

Let us now turn our attention to the dispersion characteristics of cylindrical wave-guides. When we considered the simple planar waveguide, we saw the propagating signals suffered from modal, material and waveguide dispersion. In that particular instance, we used both the simple ray model and the more complete modal analysis. Unfortunately, now that we are considering the cylindrical waveguide, we must use modal analysis and so the mathematics becomes rather involved. Thus, this section can be neglected on a first reading. In common with the previous section, all important results will be quoted later when required.

In Sect. 2.2.3, Eq. (2.78), we defined a normalised propagation constant, b, as

$$b = \frac{\beta^2 - \beta_2{}^2}{\beta_1{}^2 - \beta_2{}^2} \tag{2.118}$$

where $\beta_2 = 2\pi n_2/\lambda_o$, $\beta_1 = 2\pi n_1/\lambda_o$ and β is the propagation constant of any particular mode. If we proceed in a similar fashion to that used in Sect. 2.2.5, we can express β as

$$\beta \approx \beta_2(1 + \delta nb) \tag{2.119}$$

where we have used the weakly guiding approximation that $n_1 \approx n_2$.

Now, propagating modes travel at their own group velocities given by c/N_g. As we saw in Sect. 2.2.6, N_g can be expressed as (Eq. 2.94)

$$N_g = \frac{d\beta_2}{dk} + \frac{d}{dk}(\beta_2 b\delta n)$$
$$= N_{g2} + N_{g2}b\delta n + \beta_2 \frac{d}{dk}(b\delta n) \tag{2.120}$$

If we ignore the last term in (2.120) because it is small in comparison with the other terms, we get

$$N_g = N_{g2}(1 + b\delta n)$$

and so the group velocity is given by

$$v_g = \frac{c}{N_{g2}(1 + b\delta n)}$$

Thus the transit time of any particular mode is

$$\tau = \frac{1}{v_g} = \frac{N_{g2}}{c}(1 + b\delta n)$$

from which the modal dispersion is given by

$$\sigma_{\text{mod}} = \frac{\delta n N_{g2}}{c} \tag{2.121}$$

This expression, and derivation, is identical to that obtained for the planar optical waveguide (Eq. 2.92). We should note that this equivalence implies that only TE and TM modes propagate in the optical fibre. This is a consequence of the weakly guiding approximation, and so (2.121) is only an approximation.

In order to find the *material dispersion*, we need to find the variation of group velocity with wavelength. This is equivalent to differentiating (2.121) with respect to wavelength. So

$$D_{\text{mat}} = \frac{d}{d\lambda} \frac{\delta n N_{g2}}{c}$$

After some simple but lengthy manipulation, we find that D_{mat} is given by

$$D_{\text{mat}} = \frac{dn_1}{d\lambda}\frac{N_{g2}}{cn_2} - \frac{n_1}{n_2{}^2}\frac{N_{g2}}{c}\frac{dn_2}{d\lambda} - \lambda_o\frac{\delta n}{c}\frac{d^2n_2}{d\lambda^2}$$
$$\approx -\lambda_o\frac{\delta n}{c}\frac{d^2n_2}{d\lambda^2} \tag{2.122}$$

(This should be compared to the expression for material dispersion in a block of glass given by Eq. 2.30.)

In order to study waveguide dispersion, we can proceed in an identical manner to that used when we studied the planar optical waveguide. So, we can express the group refractive index as

$$N_g = N_{g2}\left[1 + \delta n\frac{d}{dV}(Vb)\right] \tag{2.123}$$

The first term in this expression is related to the material dispersion, while the second gives rise to the waveguide dispersion, σ_{wg}. In order to find the amount of waveguide dispersion, we need to differentiate the last term in (2.123) with respect to wavelength and then multiply by the linewidth of the source. So

$$\begin{aligned}D_{\text{wg}} &= \frac{1}{c}\frac{d}{d\lambda}\left[N_{g2}\delta n\frac{d}{dV}(Vb)\right]\\ &= \frac{1}{c}N_{g2}\delta n\frac{d}{dV}\frac{dV}{d\lambda}\frac{d}{dV}(Vb)\\ &= -\frac{N_{g2}\delta n}{c\lambda}V\frac{d^2}{dV^2}(Vb)\end{aligned} \tag{2.124}$$

where we have neglected the variation of N_{g2} and δn with wavelength. (If we had not made this assumption, the derivation would have yielded extra material dispersion terms.)

Examination of (2.124) shows that waveguide dispersion is caused by the waveguide propagation constants varying with the V value of the waveguide. At this point we could plot $d(Vb)/db$ and get the second differential by graphical means. However, in multimode fibres the waveguide dispersion is generally small when compared with modal dispersion, and so we can neglect it. When we come to consider single-mode fibres, we will find that the waveguide dispersion is of the same order of magnitude as the material dispersion and cannot be neglected.

As well as material and waveguide dispersion, there is a further source of dispersion – polarisation mode dispersion (PMD). This is of great importance in long-haul links using single-mode fibres. Basically, a mode propagating in single-mode fibre can have two states of polarisation at right angles to each other; we can designate them as E_x and E_y. Dispersion occurs when the phase coefficients associated with these signals, β_x and β_y, vary due to stresses in the fibre deforming the ideal circular cross-sectional area. This has the effect of changing the group refractive index and hence velocity for the two states of polarisation. This is known as *birefringence.* As each polarisation state carries the signal, any differential change in group velocity will cause a time difference hence dispersion.

The PMD is related to the root of the fibre length, L in km, by the PMD coefficient, D_{PMD}, as

$$\sigma_{PMD} = D_{PMD}\sqrt{L} \qquad (2.125)$$

Values of D_{PMD} vary considerably and are dependent on the fibre installation and the type of fibre. For modern fibre, D_{PMD} is 0.01 ps/√km, whereas for older fibre, it can be as high as 10 ps/√km. These figures are very small, but it should be remembered that link length can be very high particularly if erbium-doped fibre amplifiers are used instead of full regenerators and data rates are high leading to a small pulse time (40 Gbit/s gives a pulse width of 25 ps). There are some fibres that are polarisation preserving and these are useful for long-haul, high-data-rate links.

2.3.3 Step-Index Multimode Fibre

In Sect. 2.3.1 we solved Maxwell's equations in a cylindrical waveguide. We found that the fibre can support transverse electric, TE; transverse magnetic, TM; and hybrid modes, EH or HE. Here we will apply the results from the previous analysis to a certain type of fibre which is in common use today. (Although most of the parameters used in the following have already been defined, some readers may have omitted the previous section and so they are defined again in this section.)

If the refractive index of the core is very nearly that of the fibre, that is, $n_1 \approx n_2$, the fibre is known as *weakly guiding*, and it is this type of fibre which is most commonly used in telecommunications links. (Weakly guiding fibres support a small number of modes, and so modal dispersion is reduced.) With this restriction, the eigenvalue Eq. (2.117) for a fibre with radius a reduces to

$$\frac{J_{v-1}(ua)}{J_v(ua)} = -\frac{w}{u}\frac{K_{v-1}(wa)}{K_v(wa)} \qquad (2.126)$$

where $J_v(ua)$ is the Bessel function of the first kind, $K_v(wa)$ is the modified Bessel function of the second kind and v is the Bessel function order. In obtaining (2.126), the following relationships were used:

$$J'_v(ua) = -J_{v-1}(ua) + \frac{v}{ua}J_v(ua)$$

and

$$K'_v(wa) = K_{v-1}(wa) - \frac{v}{wa}K(wa)$$

The parameters u and w are defined by

$$u^2 = \left[\frac{2\pi n_1}{\lambda_o}\right]^2 - \beta^2 \text{ and } w^2 = \beta^2 - \left[\frac{2\pi n_2}{\lambda_o}\right]^2 \qquad (2.127)$$

where β is the phase constant of a particular mode. (It should be noted that (2.126) is an approximation in that it, indirectly, predicts that only TE or TM modes propagate in the fibre.)

We can define a normalised frequency variable, similar to that used with the planar waveguide, as

$$V = \frac{2\pi a\sqrt{n_1^2 - n_2^2}}{\lambda_o} \qquad (2.128)$$

From which it is easy to show that

$$V^2 = (ua)^2 + (wa)^2 \qquad (2.129)$$

As with the eigenvalue equation we encountered in Sect. 2.2.2, the solution of Eq. (2.126) involves graphical techniques. The solutions can be obtained by plotting a graph of the left- and right-hand sides of (2.126) against ua and then finding the points of intersection – the solutions to (2.126). Unfortunately, we would have to plot graphs for each value of Bessel function order, v, and for each of these plots there will be a certain number of solutions, m. Thus the propagating modes are usually known as LP_{vm} where LP refers to *linear polarisation*.

Figure 2.13 shows a plot of both sides of (2.126) for a fibre with a normalised frequency of 12.5. The curves that follow a tangent function are the left-hand side of (2.126). Two plots of the right-hand side have been drawn – the upper plot is for a Bessel function order of 0, while the lower plot is for $v = 1$. (Here we have made use of $J_{-1}(x) = -J_1(x)$ and $K_{-1}(x) = K_1(x)$.)

An interesting feature of these plots is that there are no eigenvalues for $ua > V$. Thus V is sometimes known as the *normalised cut-off frequency*. From Fig. 2.13, we can see that, provided the argument ua is less than V, the number of eigenvalues, or the number of modes, is one greater than the number of zeros for the particular Bessel function order.

Thus for the zero-order function, LP_{0m}, the number of zeros with $ua < V$ is 4, and so the number of modes is 5. For $v = 1$, LP_{1m}, the number of modes is 4; note that $ua = 0$ is a possible solution. We could find the total number of modes using this method. However, for a large V, this method would be tedious to say the least. An alternative method of estimating the number of modes is based upon a knowledge of the numerical aperture.

Fig. 2.13 Eigenvalue graphs for the zero- and first-order modes in a cylindrical waveguide

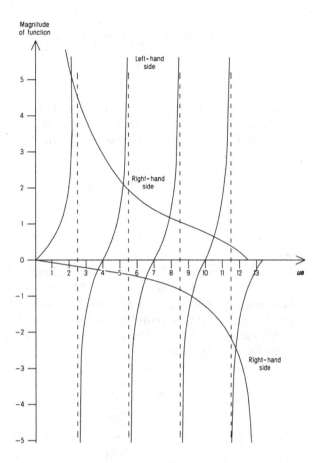

If we ignore skew rays, then the numerical aperture of a fibre will be identical to that of the planar dielectric given by Eq. (2.98). With a cylindrical fibre however, any light falling within an *acceptance cone* will propagate. The solid acceptance angle of this cone, Ω, will be given by

$$\Omega = \pi\theta_i^2 \qquad (2.130)$$

If θ_i is small, $\theta_i \approx \sin\theta_i$ which is the numerical aperture. So, Ω will be given by

$$\Omega = \pi NA^2 = \left(n_1^2 - n_2^2\right) \qquad (2.131)$$

We can now estimate the number of modes propagating by noting that the number of modes per unit angle is $2A/\lambda_o$ where A is the cross-sectional area of the fibre end. (The factor 2 is included because each mode can take on one of two different polarisation states, hence polarisation mode dispersion.) Therefore, the number of modes, N_{max}, will be

$$N_{\max} = \frac{2A}{\lambda_o}\Omega = \frac{2\pi a^2}{\lambda_o{}^2} \times \pi\left(n_1{}^2 - n_2{}^2\right)$$
$$= \frac{V^2}{2} \tag{2.132}$$

As we have already seen, evanescent waves are present in the cladding, and so not all of the transmitted power is confined to the core. By using the weakly guiding approximation, the proportion of cladding power, P_{clad}, to total power, P, can be approximated to [4]

$$\frac{P_{\text{clad}}}{P} = \frac{4}{3}\sqrt{N_{\max}} \tag{2.133}$$

Example
Light of wavelength 850 nm is propagating in 62.5 μm core diameter, PCS fibre with $n_1 = 1.5$ and $n_2 = 1.4$. The group refractive index of the cladding material, N_{g2}, is 1.53. Estimate the modal dispersion and the proportion of the total power carried in the cladding.
We can estimate the number of 850 nm modes propagating in the core by using Eq. (2.132). Thus

$$N_{\max} = \frac{V^2}{2}$$
$$= \frac{2\pi a^2}{\lambda_o{}^2} \times \pi\left(n_1{}^2 - n_2{}^2\right)$$
$$= 7.74 \times 10^3$$

As the number of propagating modes is quite large, we can approximate σ_{mod} by the expression derived for the planar waveguide, Eq. (2.92) or (2.121). Therefore

$$\sigma_{\text{mod}} = \delta n \frac{N_{g2}}{c}$$
$$= \frac{0.071 \times 1.53}{3 \times 10^8}$$
$$= 364\,\text{ps/km}$$

Such a large value of modal dispersion will tend to dominate the dispersion characteristic. Thus the bandwidth-length product of step-index, MM fibres varies from less than 1 to 100 MHz km.
As regards the distribution of power, we can use Eq. (2.133) to give

(continued)

$$\frac{P_{clad}}{P} = \frac{4}{3}\sqrt{N_{max}}$$

$$= \frac{4}{3}\sqrt{7.74 \times 10^3}$$

$$= 0.015$$

Thus only 1.5% of the total power is carried in the cladding.

2.3.4 Step-Index Single-Mode Fibre

At the start of the previous section, we saw that propagating modes had to satisfy the following eigenvalue equation

$$\frac{J_{v-1}(ua)}{J_v(ua)} = -\frac{w}{u}\frac{K_{v-1}(wa)}{K_v(wa)} \tag{2.134}$$

In a SM fibre, only the lowest-order mode can propagate. This corresponds to $V = 0$ and so (2.134) becomes

$$\frac{J_1(ua)}{J_0(ua)} = \frac{w}{u}\frac{K_1(ua)}{K_0(ua)} \tag{2.135}$$

As we saw earlier, there are m possible modes for $v = 0$. So, for SM operation, m must be equal to one (i.e. only the LP_{01} mode can propagate), and this sets a limit to ua. As the maximum value of ua is the normalised cut-off frequency, there will also be a limit to V. Now, the first discontinuity in the zero-order plot drawn in Fig. 2.13 occurs at $ua = 2.405$, and so V must be less than 2.405 for SM operation. If we use the definition of V (Eq. 2.128), then the condition for SM operation is

$$\lambda_o > 2.6a\sqrt{n_1{}^2 - n_2{}^2}$$

The term under the square root is the numerical aperture, which for practical SM fibres is usually about 0.1. Thus, for operation at 1.3 μm, the fibre diameter should be less than 10 μm, while for 1.55 μm, it should be less than 12 μm. It is interesting to note that because the condition for SM operation is dependent on wavelength, the linewidth of the source causes waveguide dispersion.

Dispersion in SM fibres is due to material and waveguide effects (as well as polarisation mode dispersion). As we saw in Sect. 2.3.2, the material dispersion in a weakly guiding optical fibre (2.122) is given by

$$D_{mat} \approx -\lambda_o \frac{\delta n}{c} \frac{d^2 n_2}{d\lambda^2} \qquad (2.136)$$

whereas the waveguide dispersion (2.124) is given by

$$D_{wg} = -\frac{N_{g2}\delta n}{c\lambda_o} V \frac{d^2}{dV^2}(Vb) \qquad (2.137)$$

In order to find D_{wg}, we need to find $d^2(Vb)/dV^2$. D. Gloge [5] has derived an expression for $d(Vb)/dV$ as

$$\begin{aligned} \frac{d(Vb)}{dV} &= b\left[1 - \frac{2J_v^2(ua)}{J_{v+1}(ua)J_{v-1}(ua)}\right] \\ &= b\left[1 - \frac{2J_0^2(ua)}{J_1^2(ua)}\right] \end{aligned} \qquad (2.138)$$

for SM operation. The argument of the Bessel functions in (2.138) is defined by Eq. (2.127). Unfortunately the second differential can only be obtained by graphical means.

Figure 2.14 shows the variation of $Vd^2(Vb)/dV^2$ with V. Most SM fibres are fabricated with V values between 2.0 and 2.4. Over this range of V, the second differential term can be approximated by

$$\frac{Vd^2(Vb)}{dV^2} \approx -\frac{1.984}{V^2} \qquad (2.139)$$

and so (2.137) can be approximated to

Fig. 2.14 Variation of b, d $(Vb)/dV$ and $Vd^2(Vb)/dV^2$ with normalised frequency, V, in an optical fibre

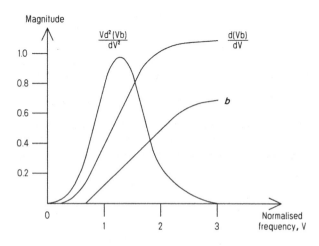

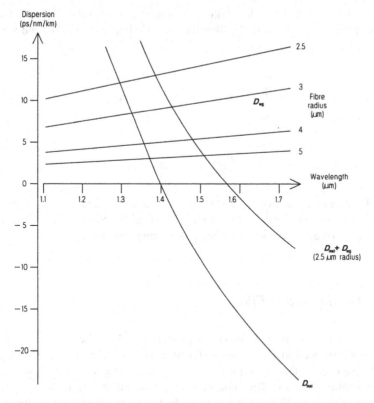

Fig. 2.15 Showing the shift in the zero dispersion point, obtained by balancing D_{mat} with D_{wg}

$$D_{wg} = \frac{N_{g2}\delta n}{c\lambda_o} \frac{1.984}{V^2} \tag{2.140}$$

In common with the material dispersion, the units of D_{wg} are usually ps/nm/km, and so we can reduce D_{wg} by using narrow linewidth sources.

In order to find the total dispersion, we can simply add together the waveguide and material dispersions. However, D_{wg} is positive, while D_{mat} becomes negative for wavelengths above about 1.3 μm. Thus the material and waveguide dispersion will cancel each other out at a certain wavelength. Figure 2.15 shows the theoretical variation with wavelength of D_{wg}, D_{mat} and total dispersion ($D_{mat} + D_{wg}$) for a typical SM fibre. As can be seen, reduction of the core radius moves the dispersion zero to higher wavelengths. (In practice, the zero dispersion point is usually limited to 1.55 μm. This is because it is difficult to manufacture very small core fibres.) Fibres which exhibit this characteristic are known as *dispersion-shifted* fibres. As the higher wavelength results in lower attenuation, they are of great importance in long-haul, high-data-rate routes. Single-mode fibres have a very high information capacity, with a typical bandwidth-length product greater than 100 GHz km.

Let us now consider the distribution of optical power in the fibre. As there is only one mode propagating in the fibre, Eq. (2.133) becomes invalid, and we must use [4]

$$\frac{P_{\text{core}}}{P} = \left(1 - \frac{u^2}{V^2}\right)\left[1 + \frac{J_0(ua)}{J_1^{\,2}(ua)}\right] \tag{2.141}$$

and

$$\frac{P_{\text{clad}}}{P} = 1 - \frac{P_{\text{core}}}{P} \tag{2.142}$$

Application of these two equations indicates that as V reduces, more power is carried in the cladding until, in the limit as $V \to 0$, all the power is in the evanescent wave. This is of great importance when considering fibre couplers.

2.3.5   Graded-Index Fibre

We have already seen that modal dispersion causes pulse distortion in MM fibres. However, we can use *graded-index* fibres to reduce this effect. The principle behind these fibres is that the refractive index is steadily reduced as the distance from the core centre increases. Thus constant refraction will constrain the propagating rays to the fibre core. With such a profile, the higher-order modes travelling in the outer regions of the core will travel faster than the lower-order modes travelling in the high refractive index region. If the index profile is carefully controlled, then the transit times of the individual modes should be identical, so eliminating modal dispersion.

The ideal index profile for these fibres is given by

$$\begin{aligned} n(r) &= n_1\sqrt{1 - 2\delta n/n_2 \times (r/a)^{\alpha}} \quad && 0 \le r \le a \\ &= n_2 \quad && r \ge a \end{aligned} \tag{2.143}$$

where n_1 is the refractive index at the centre of the core and α defines the core profile. As the wave equation is rather complex, we will not consider propagation in any detail. However, analysis shows that the optimum value of α is approximately 2. With α in this region, σ_{mod} is usually less than 100 ps/km. Of course there will still be material and waveguide dispersion effects and, depending on the source, these result in a bandwidth-length product that is typically less than 1 GHz km. (A considerable reduction in bandwidth results if α is not optimal.)

2.4   Calculation of Fibre Bandwidth

If a very narrow optical pulse propagates down a length of fibre, then, because of dispersion, the width of the output signal will be larger than that of the input. If the input pulse width is typically ten times less than the output pulse width, the output signal will closely approximate the impulse response of the fibre. Depending on the type and length of fibre, this impulse response can take on several different shapes. A Gaussian response results if there is considerable transfer of power between propagating modes – *mode mixing*. Mode mixing results from reflections off imperfections due to *micro-bending* (caused by laying the fibre over a rough surface) and scattering from fusion splices and connectors. An exponential response can also be obtained. Such an impulse response results from considerable modal dispersion in the absence of mode mixing. Of course, these are idealised extremes; in practice, the impulse response is a combination of the two. So, any bandwidth calculations performed with either pulse shape will only give an indication of the available capacity.

If we consider a Gaussian impulse response, we can write the received pulse shape, h_{out}, as

$$h_{\text{out}}(t) = \frac{1}{\sigma\sqrt{2\pi}} e^{-t^2/(2\sigma^2)} \tag{2.144}$$

Hence the received pulse spectrum, $H_{\text{out}}(\omega)$, will be given by

$$H_{\text{out}}(\omega) = e^{-\omega^2\sigma^2/2} \tag{2.145}$$

where σ is the root mean square (r.m.s.) width of the pulse. The -3 dB bandwidth is equal to the frequency at which the received power is half the d.c. power, that is

$$\frac{H_{\text{out}}(\omega)}{H_{\text{out}}(0)} = e^{-\omega^2\sigma^2/2} = \frac{1}{2} \tag{2.146}$$

and so the 3 dB bandwidth will be

$$\omega_{\text{opt}} = \frac{1.18}{\sigma} \tag{2.147}$$

We should note that this is the *optical* bandwidth. We are more usually concerned with the electrical bandwidth, that is, the bandwidth at the output of the detector. The detector converts optical power to an electrical current, and so a 3 dB drop in optical power produces a 6 dB drop in electrical power. Thus the electrical bandwidth ω_{elec} is the frequency at which the optical power is $1/\sqrt{2}$ times the d.c. value. Hence ω_{elec} will be given by

$$\omega_{elec} = \frac{0.83}{\sigma} \tag{2.148}$$

From now on, we shall use the electrical bandwidth whenever we refer to bandwidth.

The four sources of dispersion will determine the r.m.s. width of the received pulse. We have already seen that we can add the material dispersion, D_{mat}, and the waveguide dispersion, D_{wg}, together. However, we must add modal and polarisation mode dispersion together on a mean square basis. (This is a result of convolving the individual pulse shapes due to the modal and source-dependent dispersion.) So, the total dispersion, σ, will be given by

$$\sigma^2 = \sigma_{mod}{}^2 + \sigma_{pmd}{}^2 + \left(\sigma_{mat} + \sigma_{wg}\right)^2 \tag{2.149}$$

Example
An optical fibre link uses 25 μm radius, MM fibre with $n_1 = 1.5$, $N_{g1} = 1.64$, $n_2 = 1.4$, $N_{g2} = 1.53$ and $D_{mat} = 500$ ps/nm/km. An 850 nm wavelength LED, with a linewidth of 30 nm, is used as the source. Calculate the bandwidth-length product of the link assuming a Gaussian impulse response.

By applying Eq. (2.121), we get

$$\sigma_{mod} = \delta n \frac{N_{g2}}{c} = 364\,\text{ns/km}$$

We can find the material dispersion by multiplying D_{mat} by the linewidth of the source to give

$$\sigma_{mat} = 500 \times 30 = 15\,\text{ns/km}$$

The waveguide dispersion is negligible when compared with σ_{mat} and so the total fibre dispersion will be given by

$$\sigma = \sqrt{\sigma_{mod}{}^2 + \sigma_{mat}{}^2} \approx 364\,\text{ns/km}$$

Thus we can see that the modal dispersion is the dominant factor.

We can now find the bandwidth of the link by using Eq. (2.148) to give

$$\omega_{elec} = \frac{0.83}{\sigma}$$
$$= 2.3 \times 10^6\,\text{rad/s}$$

from which the bandwidth-length product is 366 kHz.km.

(continued)

Repeat the previous if the link uses SM fibre with $n_1 = 1.48$, $N_{g1} = 1.64$, $n_2 = 1.47$, $N_{g2} = 1.63$ and $D_{mat} = -5$ ps/nm/km. Assume that a 1 nm linewidth, 1.3 μm wavelength laser is used as the source. If the source is then replaced by a 30 nm linewidth LED, determine the new bandwidth.

As the fibre is single-mode, we need to find the waveguide dispersion from Eq. (2.140). Thus

$$D_{wg} = \frac{N_{g2}\delta n}{c\lambda_o} \frac{1.984}{V^2} \qquad (2.140)$$

where $V = 2.405$ for single-mode operation. Therefore

$$D_{wg} = \frac{1.63 \times 6.8 \times 10^{-3}}{3 \times 10^5 \times 1.3 \times 10^3} \times \frac{1.984}{2.405}$$

$$= 23.4 \ \text{ps/nm/km}$$

As the source has a linewidth of 1 nm, we find that

$$\sigma_{wg} = 23.4 \ \text{ps/km}$$

The material dispersion is

$$\sigma_{mat} = -5\,\text{ps/km}$$

and so the total dispersion is 18.4 ps/km. Thus the bandwidth of the link is 7.2 GHz km.

If we use a 30 nm linewidth LED source, we find

$$\sigma_{wg} = 23.4 \times 30$$

$$= 702\,\text{ps/km}$$

and

$$\sigma_{mat} = -5 \times 30$$

$$= 150 \ \text{ps/km}$$

Thus the bandwidth with the LED source is 240 MHz km.

2.5   Attenuation in Optical Fibres

Coupling losses between the source/fibre, fibre/fibre and fibre/detector can cause attenuation in optical links. Losses can also occur due to bending the fibre too far, so that the light ray hits the boundary at an angle less than θ_c. As these loss mechanisms are extrinsic in nature, we can reduce them by taking various precautions. However the fibre itself will absorb some light, and it is this attenuation that concerns us here.

The attenuation/wavelength characteristic of a typical glass fibre is shown in Fig. 2.16. This figure also shows the relative magnitudes of the four main sources of attenuation: electron absorption, Rayleigh scattering, material absorption and impurity absorption. The first three of these are known as *intrinsic absorption mechanisms* because they are a characteristic of the glass itself. Absorption by impurities is an *extrinsic absorption mechanism*, and we will examine this loss first.

2.5.1   Impurity Absorption

In ordinary glass, impurities, such as water and transition metal ions, dominate the attenuation characteristic. However, because the glass is usually thin, the attenuation is not of great concern. In optical fibres that are many kilometres long, the presence of any impurities results in very high attenuations which may render the fibre useless; a fibre made of the glass used in lenses would have a loss of several thousand dB per kilometre. By contrast, if we produced a window out of the glass used in the best optical fibres, then we would be able to see through a window 30 km thick!

The presence of water molecules can dominate the extrinsic loss. The OH bond absorbs light at a fundamental wavelength of about 2.7 μm and this, together with interactions from silicon resonances, causes harmonic peaks at 1.4 μm. 950 and 725 nm, as in Fig. 2.16. Between these peaks are regions of low attenuation – the transmission windows at 850 nm, 1.3 μm and 1.55 μm. As a high water concentration results in the tails associated with the peaks being large, it is important to minimise the OH impurity concentration.

In order to reduce attenuation to below 20 dB/km, a water concentration of less than a few parts per billion (ppb) is required (so-called dry fibre). Such values are being routinely achieved by using the *modified chemical vapour deposition* manufacturing process (examined in Sect. 2.6.2). Different manufacturing methods will produce lower water concentration. For example, the *vapour-phase axial deposition, VAD*, process can produce fibres with OH concentrations of less than 0.8 ppb. With this impurity level, the peaks and valleys in the attenuation curve are smoothed out, and this results in a typical loss of less than 0.2 dB/km in the 1.55 μm window.

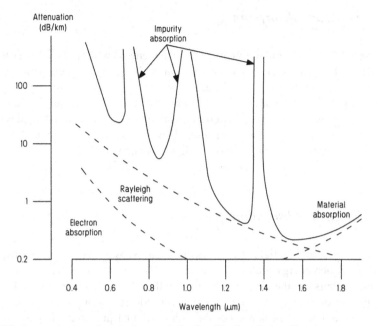

Fig. 2.16 Attenuation/wavelength characteristic of a silica-based glass fibre

The presence of transition metal ions (iron, cobalt, copper, etc.) can cause additional loss. If these metals are present in concentration of 1 ppb, then the attenuation will increase by about 1 dB/km. In telecommunications grade fibre, the loss due to transition metal ion impurities is usually insignificant when compared to the OH loss.

2.5.2   Rayleigh Scattering

Rayleigh scattering results from the scattering of light from small irregularities in the structure of the core. (A similar mechanism makes the sky appear blue, by scattering light off dust particles in the atmosphere.) These irregularities are usually due to density fluctuations which were frozen into the glass at manufacture. Consequently, this is a fundamental loss mechanism, which places a lower bound on the fibre attenuation. Rayleigh scattering is only significant when the wavelength of the light is of the same order as the dimensions of the scattering mechanism. In practice, this loss reduces as the fourth power of wavelength, and so operation at long wavelengths is desirable.

2.5.3   Material Absorption

It might be thought that operation at longer wavelengths will produce lower losses. In principle this is correct; however, the atomic bonds associated with the core material will absorb the long wavelength light – *material absorption*. Although the fundamental wavelengths of the absorption bonds are outside the range of interests, the tails are significant. Thus, operation at wavelengths greater than 1.55 μm will not produce a significant drop in attenuation. However, fibres made out of fluoride glasses, for example, ZrF_4, will transmit higher wavelength light.

2.5.4   Electron Absorption

In the ultra-violet region, light is absorbed by photons exciting the electrons in a core atom to a higher energy state. (Although this is a form of material absorption, interaction occurs on the atomic scale rather than the molecular scale.) In silica fibres, the absorption peak occurs in the ultra-violet region at about 0.14 μm; however, the tail of this peak extends through to about 1 μm, so causing attenuation in the transmission windows.

2.5.5   PCS and All-Plastic Fibres

In PCS fibre, the main absorption peaks are due to the O–H bond resonances, identical to an all-glass fibre, and the C–H bond resonances due to the plastic cladding. The net result is that PCS fibres exhibit a transmission window at 870 nm with a typical attenuation of 8 dB/km, and so PCS links can use relatively cheap near-infrared light sources. In view of the relatively low attenuation, and the fact that PCS fibre is step-index MM, most PCS links are dispersion limited rather than attenuation limited.

All-plastic fibres exhibit very high attenuation due to the presence of C–H bonds in the core material. These bonds result in a transmission window at 670 nm, with a typical attenuation of 200 dB/km. As well as the high attenuation peaks caused by the complex C–H bonds, there is a large amount of Rayleigh scattering in all-plastic fibres. This is due to scattering from the large chain molecules that make up the material. Although plastic fibres exhibit very low bandwidth-length products and very high attenuation, there is considerable interest in using such fibres for localised distribution systems such as computer installations.

2.6 Fibre Materials and Fabrication Methods

2.6.1 Materials

Most of the glass fibres in use today are fabricated out of silica, SiO_2. This has a refractive index of between 1.44 and 1.46, and doping with various chemicals produces glasses of different refractive indices. In order to increase the refractive index, oxides of germanium, GeO_2, or phosphorus, P_2O_5, are commonly used. A decrease in n results from doping with boron oxide, B_2O_3, or fluorine, F. The amount of dopant used determines the refractive index of the fibre. For example, a 5% concentration of GeO_2 will increase the refractive index of SiO_2 from 1.46 to 1.465. It should be noted that heavy doping is undesirable because it can affect both the fibre dispersion and attenuation.

Plastic clad silica, PCS, fibres are commonly made from a pure silica core with a silicone resin cladding. This gives a cladding refractive index of 1.4 at 850 nm, resulting in an acceptance angle of 20°. We can increase the NA by using a Teflon cladding. This material has a refractive index of about 1.3, resulting in an acceptance angle of 70°. As we have seen, the attenuation of these fibres is not as large as for the all-plastic fibres, and so PCS fibres find many applications in medium-haul routes.

All-plastic fibres are commonly made with a polystyrene core, $n_1 = 1.6$, and a methyl methacrylate cladding, $n_2 = 1.5$. These fibres usually have a core radius of 300 μm or more and so can couple large amounts of power. Unfortunately, because the attenuation is very high and the bandwidth very low, these fibres are only useful in very short communication links, or medical applications.

2.6.2 Modified Chemical Vapour Deposition (MCVD)

Most low-loss fibres are made by producing a glass *preform* which has the refractive index profile of the final fibre, that is, MM, SM or graded-index, but is considerably larger. If the preform is heated and a thin strand is pulled from it, then an optical fibre can be drawn from the preform. The next section describes this process in greater detail; here we will consider preform fabrication.

MCVD is probably the most common way of producing a preform. The first step in the process is to produce a SiO_2 tube, or *substrate*. This forms the cladding of the final fibre and so it may need to be doped when formed. As shown in Fig. 2.17a, the substrate is made by depositing a layer of SiO_2 particles and dopants, called a *soot*, onto a rotating ceramic former, or *mandrel*. When the soot reaches the required depth, it is vitrified into a clear glass by heating to about 1400 °C. The mandrel can then be withdrawn. (A complete preform can also be made by depositing the core glass first and then depositing the cladding. The mandrel can then be withdrawn, and the resulting tube collapsed to form a preform. This process is known as *outside vapour-phase oxidation*, and the first optical fibres with attenuations of less than 20 dB/km were made using this process.)

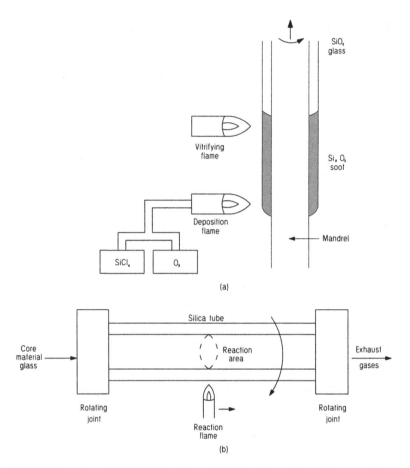

Fig. 2.17 (**a**) Formation of silica cladding tube and (**b**) deposition of core glasses

In the MCVD process, the cladding tube is placed in a lathe, and the gaseous core constituents pass through it (Fig. 2.17b). As the deposit forms, an oxyhydrogen torch sinters the core particles into a clear glass. When the required core depth is achieved, the vapour is shut off, and strong heating causes the tube to collapse. The result is a preform with the required refractive index profile. (A graded-index preform can be produced by varying the dopant concentrations during deposition.) The preform is then placed in a *pulling tower* which draws out the fibre.

2.6.3   Fibre Drawing from a Perform

Having produced a preform, the fibre is drawn from it in a *fibre pulling tower*, shown schematically in Fig. 2.18. A clamp at the top of the tower holds the preform in place, and a circular drawing furnace softens the tip.

Fig. 2.18 Schematic of a
fibre pulling tower

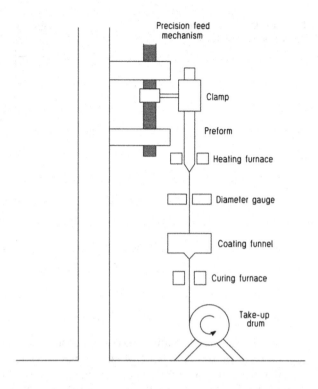

A filament of glass is drawn from the tip and attached to a take-up drum at the base of the tower. As the drum rotates, it pulls the fibre from the preform. The rate of drum rotation determines the thickness of the fibre and so a non-contact thickness gauge regulates the drum speed by means of a feedback loop.

Below the gauge, the fibre passes through a funnel containing a plastic coating which helps to protect the fibre from impurities and structural damage. A curing lamp ensures that the coating is a solid before the fibre reaches the take-up drum. A typical preform with a diameter of 2 cm, and a length of 1 m, will produce several kilometres of 125-μm-diameter fibre.

2.6.4   Fibre Drawing from a Double Crucible

A major disadvantage of fibre pulling from a preform is that the process does not lend itself to continuous production. However, if the fibre can be drawn directly from the core and cladding glasses, then a continuous process results, making the fibre cheaper to produce. Such a process is the *double crucible* method of fibre manufacture (also known as the *direct melt* technique).

Fig. 2.19 Double crucible
method of optical fibre
production

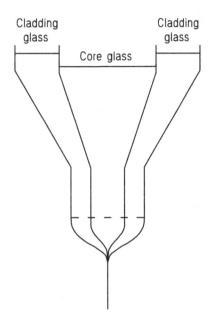

In a *double crucible* pulling tower, two concentric funnels, the double crucible, replace the preform. As Fig. 2.19 shows, the outer funnel contains the cladding material, while the inner funnel contains the core glass. In order to reduce contamination, the crucibles are usually made of platinum. The crucibles are heated to melt the glasses, and the fibre can then be drawn as previously described. Rods of the core and cladding material can be made by melting mixtures of the purified glass constituents, and a continuous drawing process results from feeding these into the crucibles. It should be obvious that this method of manufacture is only suitable for the production of step-index glass, PCS or all-plastic fibres.

2.7 Connectors and Couplers

2.7.1 Optical Fibre Connectors

When we wish to join two optical fibres together, we must use some form of connector. We could simply butt the two fibre ends together and use an epoxy resin to hold them in place. However, if the fibres move slightly while the epoxy is setting, then a considerable amount of power can be lost. One solution to the problem is to fuse the two fibre ends together, so making a stable, low-loss joint. This method, known as *fusion splicing*, is shown in schematic form in Fig. 2.20.

Fig. 2.20 Schematic
diagram of a fusion splicer

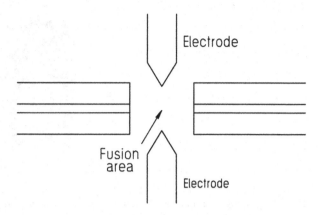

The two fibre ends are viewed through a microscope and butted together using micro-positioners. When they are correctly aligned, an electric arc is struck across the join, causing the two ends to melt and fuse. Inspection with a microscope reveals whether the joint is satisfactory; if it is not, then the join can be broken and remade. This technique results in a typical loss per splice of 0.2 dB, and so it is particularly attractive for use in long-haul routes.

Although fusion splicing results in very-low-loss connections, it does produce a permanent connection. In medium- and short-haul routes, where it may be desired to change the network configuration at some time, this is a positive disadvantage. In these systems, demountable connectors are used. There are many different types currently available, but nearly all use a precision-made ferrule to accurately align the fibre cores and so reduce losses. This method is shown in Fig. 2.21.

Prior to insertion into the connector, the protective fibre coating is first stripped off using a solvent. A taper at the end of the connector ferrule grips the inserted fibre, which usually protrudes slightly from the end. The fibre is then *cleaved* to produce a plane surface. (Cleaving involves scoring the fibre surface with a diamond and then gently bending away from the scratch until the fibre snaps. The result should be a plane end.) Any irregularities on the surface of the fibre end will scatter the light, resulting in a loss of power; thus polishing of the fibre end with successively finer abrasives is often used. The main body of the connector is then crimped on to the fibre, resulting in a mechanically strong connection. Most manufacturers will supply sources and detectors in packages which are compatible with the fibre connectors, and so installation costs can be kept low.

2.7.2 Optical Fibre Couplers

In order to distribute or combine optical signals, we must use some form of coupler. Again there are many types, but probably the most common one for use in MM systems is the Y-coupler. These can be made by butting together the chamfered ends

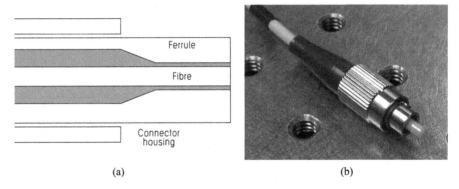

(a)   (b)

Fig. 2.21 (**a**) Basic construction of a ferrule-type connector and (**b**) an FC connector

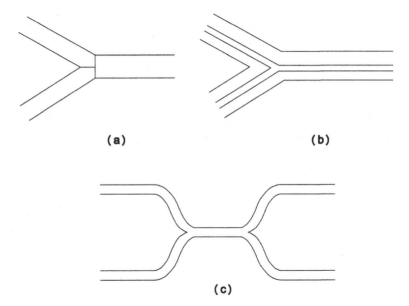

(a)   **(b)**

(c)

Fig. 2.22 (**a**) Chamfered ends of input/output fibres for (**b**) a fused *Y*-coupler (**c**) a fused biconical taper coupler

of two output fibres (Fig. 2.22a) and then fusing them with the input fibre (Fig. 2.22b). The amount of optical power sent down each arm can be controlled by altering the input fibre core area seen by each output arm.

An alternative design, which allows for multi-way splitting, is the *fused biconical taper coupler* shown in Fig. 2.22c. In this design, the fibres are first ground, or etched, to reduce the cladding thickness, twisted together and then fused, by heating to 1500 °C, to produce an interaction region. Using this basic technique, any number of fibres can be coupled together to form a *star coupler*. Couplers are commonly

Fig. 2.23 Schematic of an evanescent wave coupler

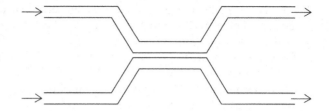

supplied with bare fibre ends, for fusion splicing, or in a package with bulkhead connectors.

Single-mode couplers rely on the coupling of the evanescent field we examined in Sect. 2.2.1. The amount of power in this field is highly dependent on the normalised frequency variable, V – a low value of V leads to a high evanescent field. The most common type of coupler is the fused biconical taper we have just discussed. As the amount of coupling is dependent upon the contact length and cladding thickness, the fibres are stretched while being heated. This stretching reduces the core diameter, and so the value of V falls. This has the effect of increasing the power in the evanescent field, so increasing the coupling. It should be noted that power can be coupled from, and to, either fibre.

An alternative coupler can be made by implanting a dielectric waveguide into a substrate (Fig. 2.23). The most commonly used substrate material is *lithium niobate*, $LiNbO_3$. The guides are made by diffusing titanium into the substrate. In common with the SM fibre coupler, power is transferred between the waveguides through the evanescent fields. Couplers of this type form the basis of a large family of components known as *integrated optics*.

An important parameter to be considered when specifying couplers is the insertion loss, or *excess loss*. This is the ratio of the total output power to the input power. Typical MM couplers have an excess loss of 3 dB, while that of SM couplers can be less than 0.5 dB. The major source of loss in couplers is the attenuation introduced by the connections, and so it is important to use low-loss connectors or fusion splices.

Problems

1. Light of wavelength 1.2 μm is propagating through a block of glass with $\mu_r = 1$, $\varepsilon_r = 5$. Determine the refractive index, the phase velocity and the phase coefficient.
   [2.24, 1.34×10^8 m/s, 11.71×10^6 rad/m]
2. Determine the average power collected in an area of 1 mm^2 if the incident light has an r.m.s. electric field strength of 1 kV/m in air.
   [2.65 mW]
3. Light of wavelength 1.5 μm is propagating in a 5 μm thickness planar waveguide. If the inner waveguide has a refractive index of 1.5 and is surrounded by material with refractive index 1.45, determine the number of modes and their angles of incidence.

   [6 modes at 87.42°, 84.87°, 82.35°, 79.85°, 77.40°, 75.25°]

4. For the waveguide of Q3, determine the numerical aperture and the acceptance angle assuming the waveguide is in air.

[0.38, 22.60°]

5. Determine the numerical aperture of a step-index fibre with core diameter 50 μm, a core refractive index of 1.5 and a cladding refractive index of 1.4. Compare to that of a single-mode fibre with core diameter of 4 μm, a core refractive index of 1.495 and a cladding refractive index of 1.5.

[0.54, 0.12]

6. Calculate the V number and estimate the number of propagating modes for the fibres of Q5. Take an operating wavelength of 850 nm and 1.55 μm for the multimode and the single-mode fibre, respectively. If the multimode fibre has a group refractive index in the cladding of 1.53, estimate the modal dispersion and the fibre bandwidth.

[99.5, 0.99, 4980, 1, 364 ns/km, 363 MHz.m]

7. A single-mode fibre has waveguide dispersion of 10 ps/(nm.km) and material dispersion of −8 ps/(nm.km). In addition there is polarisation mode dispersion of 0.1 ps/√km. Determine the total dispersion and the bandwidth for fibre lengths of 10 km, 50 km and 100 km. Assume a source linewidth of 0.01 nm.

[0.37 ps, 425 GHz; 1.2 ps, 130 GHz; 2.2 ps, 71 GHz]

Recommended Readings

1. Fleisch D (2008) A student's guide to Maxwell's equations. Cambridge University Press, Cambridge, UK
2. Fleisch D (2015) A student's guide to waves. Cambridge University Press, Cambridge, UK
3. Agrawal GP (2010) Fiber-optic communication systems. Wiley, Hoboken, NJ
4. Gloge D (1971) Weakly guiding fibres. Applied Optics 10:2252–2258
5. Ainslie BJ et al (1982) Monomode fibre with ultralow loss and minimum dispersion at 1.55 μm. Electronics Letters 18:843–844

Chapter 3
Optical Transmitters

To be useful in an optical link, a light source needs the following characteristics:

1. It must be possible to operate the device continuously at a variety of temperatures for many years.
2. It must be possible to modulate the light output over a wide range of modulating frequencies.
3. For fibre links, the wavelength of the output should coincide with one of the transmission windows for the fibre type used.
4. To couple large amounts of power into an optical fibre, the emitting area should be small.
5. To reduce material dispersion in an optical fibre link, the output spectrum should be narrow.

We shall examine several sources that satisfy these requirements – the light-emitting diode, *LED*; the semiconductor laser diode, *SLD;* solid-state and gas lasers; and fibre lasers. Before we examine these sources in greater detail, it will be useful to discuss light emission in semiconductors and semiconductor physics in general.

3.1 Semiconductor Diodes

All *semiconductor* light sources use a forward biased p-n junction diode to generate light. As LEDs and SLDs are commonly used in optical fibre links, it will be useful for us to examine the physics of a semiconductor p-n junction in some detail. Let us begin our study by examining intrinsic semiconductor material, that is, material that has not been doped to either p- or n-type.

© Springer Nature Switzerland AG 2020

M. Sibley, *Optical Communications*, https://doi.org/10.1007/978-3-030-34359-0_3

3.1.1 Intrinsic Semiconductor Material

Under thermal equilibrium, there will be a certain number of electrons available for conduction in the conduction band, *CB*, and a corresponding number of holes in the valence band *VB*. The density of free electrons in the CB is generally quoted in terms of the *Fermi-Dirac probability function*. This function describes the most likely distribution of electron energies as

$$F(E) = \frac{1}{1 + \exp\left[(E - E_f)/kT\right]} \tag{3.1}$$

where F (E) is the probability that an electron has an energy E and E_f is the energy level at which $F(E)$ is exactly 0.5 - the *Fermi level*. We can simplify (3.1) by noting that, in the conduction band, $E - E_f$ is generally greater than kT and so

$$F(E) \approx \exp\left[-\frac{E - E_f}{kT}\right] \tag{3.2}$$

We can check (3.2) by noting that as temperature increases, $F(E)$ tends to unity and so all the electrons are thermally excited to the CB. Conversely, if the temperature tends to absolute zero, then $F(E)$ tends to zero and there are no electrons available for conduction.

In order to find the density of thermally excited electrons in the CB, n, we simply multiply the density of available levels, $S(E)$, by the probability of finding an electron in a particular level and integrate over all available levels. So

$$n = \int_{E_c}^{E_t} S(E)F(E) \tag{3.3}$$

where E_c is the energy level at the bottom of the CB and E_t is the level at the top of the CB. As (3.3) tends to zero very quickly as E tends to E_t, we can take the upper level in (3.3) to be ∞ to a good approximation. Thus the integral in (3.3) results in

$$n \approx N_c \exp\left[-(E_c - E_f)/kT\right] \tag{3.4}$$

where

$$N_c = \frac{2(2\pi m_e kT)^{3/2}}{h^2}$$

and m_e is the effective mass of an electron.

We can find the density of holes by noting that every thermally generated free electron leaves behind it a hole. Thus the probability that a level is *not filled* is 1 - F *(E)* and the density of holes is given by

$$p \approx N_v \exp\left[-(E_f - E_v)/kT\right] \qquad (3.5)$$

where

$$N_v = \frac{2(2\pi m_h kT)^{3/2}}{h^2}$$

and m_h is the effective mass of a hole.

As we have already noted, each thermally generated electron leaves behind a hole, and so the density of electrons in the CB must be the same as the density of holes in the VB, that is, $n = p$. Thus

$$N_c \exp\left[-(E_c - E_f)/kT\right] = N_v \exp\left[-(E_f - E_v)/kT\right]$$

which, after some rearranging, becomes

$$E_f = \frac{E_g}{2} + \frac{kT}{2} \ln\left(N_v/N_c\right) \qquad (3.6)$$

where E_g is the band-gap of the semiconductor material. As m_h and m_e are of the same order of magnitude, we can approximate $\ln(N_v/N_c)$ to zero. Thus we can see that the Fermi level in intrinsic semiconductor lies mid-way between the VB and CB as shown in Fig. 3.1. We can also find the density of carriers in intrinsic material, n_i, by noting

$$
\begin{aligned}
n_i^2 &= np \\
&= N_c \exp\left[-(E_c - E_f)/kT\right] N_v \exp\left[-(E_f - E_v)/kT\right] \\
&= N_c N_v \exp\left[-(E_c - E_v)/kT\right] \\
&= N_c N_v \exp\left[-E_g/kT\right]
\end{aligned}
$$

or

$$n_i = \sqrt{N_c N_v} \exp\left[-E_g/2kT\right] \qquad (3.7)$$

We should note that (3.7) is independent of E_f and so it will apply equally well to intrinsic and extrinsic semiconductors. Thus we can say that the *electron-hole product in intrinsic and extrinsic semiconductor is a constant.*

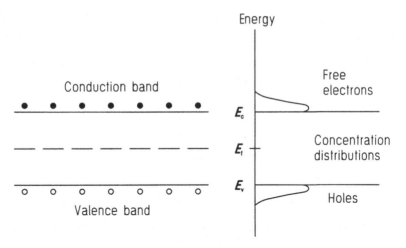

Fig. 3.1 Fermi level and carrier concentration distribution in intrinsic semiconductor material

Example

The effective mass of electrons in intrinsic silicon at a temperature of 300 Kelvin is 8.8×10^{-31} kg and that of holes is 4.6×10^{-31} kg. Estimate the density of electrons in the conduction band and of holes in the valence band. Also find the Fermi level and the total carrier density. The band-gap energy for silicon is 1.12 eV. Repeat for GaAs for which the effective mass of electrons and holes is 0.61×10^{-31} kg and 4.1×10^{-31} kg, respectively, and the band-gap is 1.44 eV.

We can find N_c and N_v by using

$$N_c = 2\left(\frac{2\pi m_e kT}{h^2}\right)^{3/2} \quad \text{and} \quad N_v = 2\left(\frac{2\pi m_h kT}{h^2}\right)^{3/2}$$

Thus $N_c = 2.4 \times 10^{25}$ m^{-3} and $N_v = 0.9 \times 10^{25}$ m^{-3}. The Fermi level is given by Eq. (3.6) as

$$E_f = \frac{E_g}{2} + \frac{kT}{2q} \ln\left(N_v/N_c\right)$$

$$= \frac{E_g}{2} - 0.013$$

$$\approx \frac{E_g}{2}$$

(continued)

Thus we can see that the intrinsic carrier densities, although quite high, do not significantly affect the Fermi level. (We include the electronic charge in this derivation to give the Fermi level in electronvolts.)

We can find the total carrier density from Eq. (3.7). Thus

$$n_i = \sqrt{N_c N_v} \exp\left[-E_g/2kT\right]$$

$$= \sqrt{2.4 \times 10^{25} \times 0.9 \times 10^{25}} \exp\left(\frac{-1.6 \times 10^{-19} \times 1.12}{2 \times 1.38 \times 10^{-23} \times 300}\right)$$

$$= 7.1 \times 10^{15} \text{ m}^{-3}$$

For GaAs, N_c and N_v are 4.3×10^{23} m^{-3} and 76×10^{23} m^{-3}, respectively; $E_f = E_g/2$; carrier density $= 1.46 \times 10^{12}$ m^{-3}.

3.1.2   Extrinsic Semiconductor Material

Let us now turn our attention to extrinsic semiconductor material. When we dope intrinsic material with donor atoms, the donor atom density, N_d, is usually sufficient to mask the effects of thermally generated electrons. Thus $n = N_d$ and (3.4) becomes

$$N_d \approx N_c \exp\left[-(E_c - E_{fn})/kT\right] \tag{3.8}$$

from which the Fermi level is given by

$$E_{fn} = E_c + kT \ln\left(N_d/N_c\right) \tag{3.9}$$

If we dope the material with acceptor atoms, we get

$$N_a \approx N_v \exp\left[-\left(E_{fp} - E_v\right)/kT\right] \tag{3.10}$$

and

$$E_{fp} = E_v - kT \ln\left(N_a/N_v\right) \tag{3.11}$$

Thus we can see that when we dope intrinsic material with donor atoms, the Fermi level moves towards the CB, and so there is a greater probability of finding electrons in the CB, so-called *n-type* material, whereas when we dope with acceptor atoms, the Fermi level moves towards the VB, so-called *p-type* material. This situation is shown in Fig. 3.2.

If we apply a voltage to an extrinsic semiconductor so that majority carriers (electrons in n-type and holes in p-type) are injected into the material, a current will

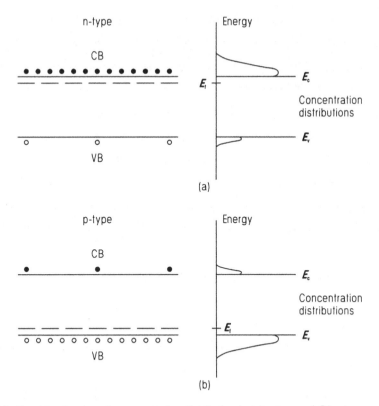

Fig. 3.2 Fermi level and carrier concentration distribution in (**a**) n-type and (**b**) p-type semiconductor material

flow under the influence of the applied bias. So, for the n-type material, the current density due to the drift of injected carriers (electrons) will be

$$
\begin{aligned}
J_{ndrift} &= nqv_d \\
&= nq\mu_n E
\end{aligned}
\tag{3.12}
$$

where v_d is the *drift velocity* of the electrons given by $v_d = \mu_n E$, μ_n is the *electron mobility,* and E is the *electric field strength.* We can write a similar expression for the drift current due to holes being injected into the p-type material as

$$
J_{pdrift} = pq\mu_p E
\tag{3.13}
$$

As well as the drift of carriers throughout the material, there can also be *diffusion* of *minority* carriers down a concentration gradient. Let us consider a sample of n-type material, with no applied bias. Next, let us introduce, by some means, a large number of minority carriers (holes) at the left-hand side of the sample. (The holes could appear as a result of the generation of electron-hole pairs caused by the absorption of light; see Chap. 4.) These holes will tend to diffuse down a concentration gradient, away from the area where they were produced. (A similar situation occurs with the diffusion of gases.) This gives rise to a *diffusion current* given by for holes in n-type material

$$J_{\text{diff}} = -D_p q \frac{dp_n}{dx} \tag{3.14}$$

and, for electrons in p-type material

$$J_{\text{diff}} = -D_n q \frac{dn_p}{dx} \tag{3.15}$$

where D_n and D_p are the diffusion coefficients for electrons and holes, respectively. (The negative sign in these equations arises from the fact that the minority carrier concentration gradient *reduces* as distance x increases.) As these minority carriers are in a region of high majority carrier density, they will tend to recombine as they diffuse through the sample. In particular, the minority carrier density will reach the background level after one *diffusion length* – symbol L_n for electrons in p-type and L_p for holes in n-type. (We will return to the diffusion length shortly.)

So, we have seen that the current in a block of semiconductor material can consist of drift current, due to the movement of *majority* carriers under the influence of an electric field and diffusion current due to the diffusion of *minority* carriers down a concentration gradient. As we will see in the next section, both types of current are present in a p-n junction diode.

Example
P-type GaAs is formed by doping intrinsic material with Zn atoms at a density of 10^{24} m^{-3}. Determine the Fermi level if GaAs has $E_g = 1.424$ eV, $N_c = 4.7 \times 10^{23}$ m^{-3}, and $N_v = 7 \times 10^{24}$ m^{-3}.

If the sample is 1 cm long, and a voltage of 5 volts is placed across it, determine the total current density in the sample. (The mobility of electrons and holes in GaAs is 0.85 and 0.04 m^2 / V s, respectively.)

The intrinsic carrier density due to thermal excitation is given by

$$n_i = \sqrt{N_c N_v} \exp\left[-E_g/2kT\right] = 1.46 \times 10^{12} \text{ m}^{-3}$$

and so we can see that a doping density of 1×10^{24} m^{-3} will mask the effects of thermally generated carriers. Thus the Fermi level will be given by (Eq. 3.11)

(continued)

$$E_{fp} = E_v - \frac{kT}{q} \ln (N_a/N_v)$$
$$= E_v + 0.05$$

Again, because the band-gap is quoted in electronvolts, we must include the electronic charge in the calculations.

The drift current of the electrons will be given by (3.12). So

$$J_{ndrift} = \frac{n_i^2}{N_a} q\mu_n E$$
$$= \frac{(2 \times 10^{12})^2}{1 \times 10^{24}} 1.6 \times 10^{-19} \times 0.85 \times \frac{5}{1 \times 10^{-2}}$$
$$= 2.72 \times 10^{-16} \text{ A/m}^2$$

Similarly, the drift current of the holes will be Eq. (3.13)

$$J_{pdrift} = pq\mu_p E$$
$$= 3.2 \times 10^6 \text{ A/m}^2$$

Thus we can see that the majority carrier drift current is significantly higher than the minority carrier current.

3.1.3 The p-n Junction Diode Under Zero Bias

A p-n junction diode is formed by joining p- and n-type extrinsic semiconductors together. A *heterojunction* is formed if the p- and n-type materials are different, whereas we get a *homojunction* if the materials are identical. Heterojunction diodes are dealt with later: here we will examine homojunction diodes.

Figure 3.3 relates to a p-n junction diode under zero bias. As shown in Fig. 3.3b, the hole concentration in the p-type, where they are in a majority, is far greater than the concentration in the n-type where they are in a minority. This gives rise to a concentration gradient down which the holes will diffuse. When these holes reach the n-type material, they will recombine with the free electrons, so consuming some majority carriers. A similar situation occurs with electrons diffusing into the p-type region. So, there will be an area either side of the junction that is depleted of carriers – the so-called *depletion region.*

Now, as holes migrate across the junction, they leave behind them negative acceptor atoms. (The acceptor atoms in the p-type are tightly bound in the crystal lattice, and so they cannot follow the holes.) This has the effect of making the depletion region in the p-type negatively charged. A similar situation causes a

Fig. 3.3 The p-n junction under zero bias: (**a**) schematic, (**b**) carrier distribution, (**c**) charge distribution, (**d**) variation in potential and (**e**) electric field distribution

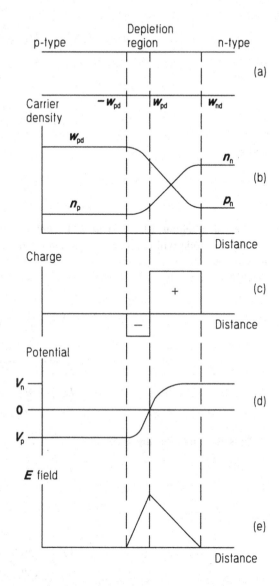

positively charged region in the n-type. Thus there is a charge distribution as shown in Fig. 3.3c. This distribution gives rise to an electric field as shown in Fig. 3.3e.

The direction of the electric field is such as to oppose the diffusion of carriers from both sides of the junction. So, in effect we have both diffusion and drift currents in the depletion region. However, as there is no bias across the diode, there can be no net flow of current and so the sum of drift and diffusion currents must equal zero.

Let us initially consider the flow of electrons across the junction. As we can reason from Fig. 3.3, the electron drift and diffusion currents act in the same direction. Thus, from (3.12) and (3.15) we get

$$J_n = -nq\mu_n E - D_n \frac{qdn_p}{dx} = 0 \tag{3.16}$$

and as $E = -dV/dx$

$$J_n = +nq\mu_n \frac{dV}{dx} - D_n \frac{qdn_p}{dx} = 0$$

and so

$$dV = \frac{D_n}{n\mu_n} dn_p \tag{3.17}$$

We can integrate (3.17) to give the *barrier potential* that must be overcome before the diode will conduct. The limits of the integral will be $V_p < V < V_n$ and $n_p < n < n_n$ where n_n is the density of electrons in the n-type and n_p the density of electrons in the p-type. So, if we integrate (3.17), we get

$$V_b = V_n - V_p = \frac{D_n}{\mu_n} \ln\left(\frac{n_n}{n_p}\right) \tag{3.18}$$

Now, $n_n = N_d$ and $n_i^2 = np = n_p N_a$, and so we can write

$$V_b = \frac{D_n}{\mu_n} \ln\left(\frac{N_d N_a}{n_i^2}\right)$$

Einstein's relationship states that

$$\frac{D_n}{\mu_n} = \frac{kT}{q}$$

and so, V_b becomes

$$V_b = \frac{kT}{q} \ln\left(\frac{N_d N_a}{n_i^2}\right) \tag{3.19}$$

Thus we can see that the barrier potential is such that the n-type material is positive with respect to the p-type. The term kT/q is approximately 25 mV at room temperature, and so the barrier voltage is dependent on the doping level in the p- and n-type material.

We can find the width of the depletion region, and hence the capacitance of the diode, by noting that the total charge in the depletion region under zero bias must equal zero. Thus we can write

$$w_{pd}N_a = w_{nd}N_d \tag{3.20}$$

where w_{pd} and w_{nd} are the widths of the depletion regions in the p-type and n-type, respectively. We can manipulate this equation to write the total width of the depletion region as

$$w = w_{nd}(1 + N_d/N_a) \tag{3.21}$$

We should note that the depletion layer will extend unequally into the n-side if the p-type is heavily doped - a p + n diode. Such diodes are often used as avalanche photodiodes, dealt with in Chap. 4.

In order to find w_{pd} and w_{nd}, we need to apply Poisson's equation:

$$\frac{d^2V}{dx^2} = -\frac{Pt}{\varepsilon} \tag{3.22}$$

where Pt is the charge density in the material and ε is the permittivity of the semiconductor. In the n-type material we can write

$$\frac{d^2V}{dx^2} = -\frac{qN_d}{\varepsilon}$$

and so

$$\frac{dV}{dx} = -\frac{qN_d x}{\varepsilon} + \text{constant} \tag{3.23}$$

We can find the value of the constant by noting that the E field is zero at the edge of the n-type depletion region. Thus

$$\frac{dV}{dx} = 0 \text{ at } x = -w_{nd}$$

and so the constant in (3.23) is

$$-\frac{qN_d}{\varepsilon}w_{nd}$$

Equation (3.23) now becomes

$$\frac{dV}{dx} = -\frac{qN_d}{\varepsilon}(w_{nd} + x)$$

One further integration yields

$$V_n = -\frac{qN_d}{\varepsilon}\left(w_{nd}x + \frac{x^2}{2}\right) + \text{constant} \qquad (3.24)$$

As we are taking the potential at the junction between the two materials to be zero, the constant in (3.24) is zero. Thus

$$V_n = -\frac{qN_d}{\varepsilon}\left(w_{nd}x + \frac{x^2}{2}\right) \qquad (3.25)$$

in the n-type material. By following a similar procedure with the p-type material, we get

$$V_p = \frac{qN_a}{\varepsilon}\left(\frac{x^2}{2} - w_{pd}x\right) \qquad (3.26)$$

Thus the barrier potential is also given by

$$\begin{aligned} V_b &= V_{n@x=-w_{nd}} - V_{p@x=w_{pd}} \\ &= \frac{q}{2\varepsilon}\left(N_d w_{nd}{}^2 + N_a w_{pd}{}^2\right) \end{aligned} \qquad (3.27)$$

In order to find the width of the depletion region, we can find V_b from (3.19) and equate it to V_b from (3.27). This is done in the example at the end of this section.

Let us now turn our attention to the variation in carrier density across the depletion region. If we consider the edge of the depletion layer in the p-type, that is $x = -w_{pd}$, we can write from (3.19)

$$\frac{qV_b}{kT} = \ln\left(\frac{N_d N_a}{n_i{}^2}\right)$$

and so

$$\frac{N_d N_a}{n_i{}^2} = \exp\left(\frac{qV_b}{kT}\right)$$

Now, $n_i{}^2 = n_p N_a$ at $x = -w_{pd}$ and so

$$\frac{N_d N_a}{n_p N_a} = \exp\left(\frac{qV_b}{kT}\right)$$

from which the density of minority carriers in the p-type is

$$n_p = N_d \exp\left(-qV_b/kT\right) \tag{3.28}$$

at the edge of the depletion region. By a similar procedure, we can write the density of holes in the n-type as

$$p_n = N_a \exp\left(-qV_b/kT\right) \tag{3.29}$$

Thus we can see that, in crossing the depletion region, the carrier densities fall from their maximum values of N_d and N_a to n_p and p_n.

Example

A silicon p + −n junction diode is formed from p-type material with $N_a = 10^{24}$ m^{-3} and n-type material with $N_d = 10^{21}$ m^{-3}. Determine the barrier potential, the width of the depletion region, the maximum field strength, and the concentration of minority carriers at the depletion region boundaries. (The density of thermally generated carriers in silicon is 1.4×10^{16} m^{-3} and $\varepsilon_r = 11.8$.)

From (3.19) the barrier potential is

$$
\begin{aligned}
V_b &= \frac{kT}{q} \ln\left(\frac{N_d N_a}{n_i^2}\right) \\
&= 25 \times 10^{-3} \times \ln\left(5.1 \times 10^{12}\right) \\
&= 0.73 \text{ volt}
\end{aligned}
$$

This is the familiar voltage drop produced by a silicon diode. In order to find the width of the depletion region, we must apply (3.27). Thus

$$V_b = \frac{q}{2\varepsilon}\left(N_d w_{nd}^2 + N_a w_{pd}^2\right)$$

and so

$$0.73 = 7.7 \times 10^{11} w_{nd}^2 + 7.7 \times 10^{14} w_{pd}^2$$

Now, as the total charge in the depletion region equals zero, we have

$$w_{pd} N_a = w_{nd} N_d$$

or

(continued)

$$w_{pd} = w_{nd} \times 10^{-3}$$

Therefore

$$0.73 = 7.7 \times 10^{11} w_{nd}{}^2 + 7.7 \times 10^8 w_{nd}{}^2$$

and so $W_{nd} = 1$ μm and $w_{pd} = 1$ nm.

Thus we can see that the width of the depletion layer is approximately 1 μm, and it is mainly in the lightly doped n-type material.

The maximum field strength occurs at the junction between the two materials. Thus

$$E_{max} = -\frac{dV}{dx} @x = 0$$

$$= \frac{qN_d}{\varepsilon} w_{nd}$$

$$= 1.5 \ \text{MV}/_m$$

$$= 1.5 \ \text{V}/\mu m$$

We can find the minority carrier densities by using (3.28) and (3.29) to give

$$n_p = 1 \times 10^{21} \exp\left(-\frac{qV_b}{kT}\right)$$

$$= 4.6 \times 10^8 \ \text{m}^{-3}$$

and

$$p_n = 1 \times 10^{24} \exp\left(-\frac{qV_b}{kT}\right)$$

$$= 4.6 \times 10^{11} \ \text{m}^{-3}$$

This example has shown that if the p- and n-type doping levels are different by three orders of magnitude, the depletion region is mainly in the lightly doped part of the semiconductor. We will return to this point when we consider photodiodes in the next chapter.

3.1.4   The p-n Junction Diode Under Forward Bias

In the previous section, we saw that a p-n junction diode has a depletion region across the junction. Under conditions of zero bias, the drift and diffusion currents balance each other out. However, if we bias the diode by connecting the n-type to a

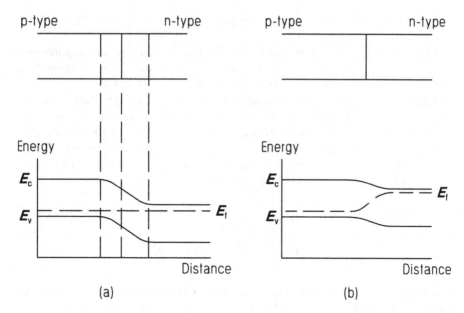

Fig. 3.4 Schematic and energy diagram for a p-n junction diode under (**a**) zero bias and (**b**) forward bias

source of electrons, the barrier potential is reduced as shown in Fig. 3.4. This has the effect of upsetting the balance between drift and diffusion currents, and a current will flow through the diode. All semiconductor light sources generate light under forward bias, and so an understanding of this area will help us in our later analyses. (Photodiodes operate under reverse bias, and so we will consider this in the next chapter.)

Under a forward bias of voltage V, the voltage across the depletion region will be $V_d = V_b - V$. Thus the minority carrier densities at the edges of the depletion region will become

$$n'_p = N_d \exp\left(-qV_b/kT\right) \exp\left(qV/kT\right)$$

and

$$p'_n = N_a \exp\left(-qV_b/kT\right) \exp\left(qV/kT\right)$$

By substituting from (3.28) and (3.29), we can write

$$n'_p = n_p \exp\left(qV/kT\right) \tag{3.30}$$

and

$$p'_n = p_n \exp{(qV/kT)} \qquad (3.31)$$

These equations show that, under forward bias, the minority carrier density on either side of the depletion region increases exponentially with the bias voltage. This excess of carriers causes a large diffusion current to flow through the diode.

Let us initially consider the component of diode current caused by the diffusion of minority carriers and ignore recombination. The minority carrier concentration in the p-type varies from a maximum of n'_p, at the edge of the depletion layer, to a minimum of n_p at the p-type contact. Similarly, in the n-type the minority carrier concentration goes from a maximum of p'_n to a minimum of p_n at the n-type contact. So, the diffusion current density will be given by

$$\begin{aligned} J &= D_n q \frac{dn}{dx} - D_p q \frac{dp}{dx} \\ &= D_n q \frac{n'_p - n_p}{x_p - w_{pd}} + D_p q \frac{p'_n - p_n}{x_n - w_{nd}} \end{aligned}$$

where x_p and x_n are the widths of the p- and n-type regions, respectively. As the diode is forward biased, the depletion region will be very small and so w_{pd} and w_{nd} will be almost zero. So, if we make this assumption, and substitute for n'_p and p'_n, we get

$$J = J_0[\exp{(qV/kT)} - 1] \qquad (3.32)$$

where

$$J_0 = D_n q \frac{n_p}{x_p} + D_p q \frac{p_n}{x_n}$$

As J is the current density, the diode current will also follow the same form as (3.32).

Let us now take account of carrier recombination. Electrons injected into the p-type will recombine with holes to try to maintain thermal equilibrium. The holes that are lost through recombination are replaced by the injection of carriers from the external contact, and so there will be an electric field across the diode. Thus the current density at any point will be made up of the diffusion of minority carriers towards the external contact, J_{ndiff}, and the drift of majority carriers, J_{hr}, from the external contact. At all points in the p-type region, the total current will be constant and given by

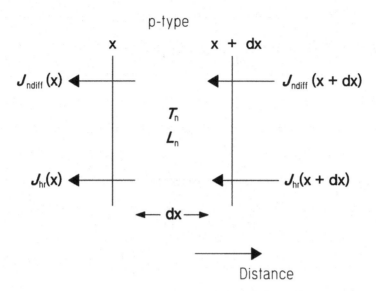

Fig. 3.5 Current flow at a certain point in the p-type region of a forward biased p-n junction diode

$$J_t = J_{ndiff} + J_{hr}$$

Figure 3.5 shows the situation at a certain point in the p-type material. Carrier recombination can be described in terms of the *carrier lifetime*, τ_n. This is defined as the time taken for an injected carrier concentration to fall to $1/e$ times its original value. Since the total current is constant across the p-type, we can write

$$\frac{d}{dx}J_t = 0$$

that is

$$\frac{d}{dx}J_{ndiff} + \frac{d}{dx}J_{hr} = 0 \qquad (3.33)$$

Now, the charge lost per second due to recombination in the element dx is

$$Q = \frac{q(n - n_p)}{\tau_n}Adx$$

The p-type contact must supply this charge and so

$$Q = dI_{hr}$$

that is

$$dI_{hr} = \frac{q(n - n_p)}{\tau_n} A dx.$$

and so

$$\frac{d}{dx} J_{hr} = \frac{q(n - n_p)}{\tau_n} \qquad (3.34)$$

As regards the diffusion current, we can write

$$\frac{d}{dx} J_{ndiff} = -\frac{d}{dx}\left(qD_n \frac{dn}{dx}\right)$$

$$= -qD_n \frac{d^2 n}{dx^2} \qquad (3.35)$$

By substituting Eqs. (3.34) and (3.35) into (3.33), we get

$$D_n \frac{d^2 n}{dx^2} - \frac{(n - n_p)}{\tau_n} = 0 \qquad (3.36)$$

In order to solve this equation, we need to apply the boundary conditions that $n = n'_p$ at $x = 0$ and $n = n_p$ at $x = -x_p$. (Here we are assuming that the width of the depletion region is negligible when compared with the diode dimensions.) So, the solution to (3.36) is

$$n(x) - n_p = \left(n'_p - n_p\right) \exp\left(\frac{-x}{\sqrt{D_n \tau_n}}\right) \qquad (3.37)$$

From Eq. (3.37) we can see that the excess electron density falls to $1/e$ times its original value in distance $\sqrt{D_n \tau_n}$. This distance is known as the *diffusion length, Ln.* By following a similar analysis with holes, we get

$$p(x) - p_n = \left(p'_n - p_n\right) \exp\left(\frac{-x}{\sqrt{D_p \tau_p}}\right) \qquad (3.38)$$

and so the diffusion length for holes is $L_p = \sqrt{D_p \tau_p}$.

Example

A GaAs p + n junction diode is formed from p-type material with $N_a = 10^{24}$ m^{-3} and n-type material with $N_d = 10^{21}$ m^{-3}. Determine the concentration of minority carriers as a function of distance from the junction, assuming an external forward bias of 1.3 volt. What is the current density if the p- and n-type are both ten diffusion lengths long? (In GaAs, $\varepsilon_r = 13.1$, $D_n = 22 \times 10^{-3}$ m^2/s, $D_p = 1 \times 10^{-3}$ m^2/s, $\tau_n = \tau_p = 50$ ns.)

By following a similar procedure to that used in the previous example, we find that the barrier potential of this diode is 1.2 volt. Thus the equilibrium carrier densities are

$$n_p = 5\,\mathrm{m}^{-3} \quad \text{and} \quad p_n = 5 \times 10^3 \mathrm{m}^{-3}$$

Now,

$$n'_p = n_p \exp\left(qV/kT\right)$$

and

$$p'_n = p_n \exp\left(qV/kT\right)$$

and so

$$n'_p = 4.62 \times 10^{22} \mathrm{m}^{-3}$$

and

$$p'_n = 4.62 \times 10^{25} \mathrm{m}^{-3}$$

Thus the excess carrier densities are

$$n(x) - n_p = \left(n'_p - n_p\right) \exp\left(\frac{-x}{\sqrt{D_n \tau_n}}\right)$$
$$= 4.62 \times 10^{22} \exp\left(-30 \times 10^3 x\right)$$

and

$$p(x) - p_n = 4.62 \times 10^{25} \exp\left(-140 \times 10^3 x\right)$$

As regards the current density, we can use (3.32) to give

(continued)

$$J = \left(D_n q \frac{n_p}{x_p} + D_p q \frac{p_n}{x_n}\right)[\exp(qV/kT) - 1]$$

Now, x_p and x_n are both ten diffusion lengths. Thus we can write

$$x_p = 10\sqrt{D_p \tau_p} \quad \text{and} \quad x_n = 10\sqrt{D_n \tau_n}$$

giving

$$x_p = 70\,\mu\text{m} \quad \text{and} \quad x_n = 330\,\mu\text{m}$$

So

$$J = 2.7 \times 10^{-15} \times 9.2 \times 10^{21}$$
$$= 2.5 \times 10^7 \, \text{A/m}^2$$

This example has shown that, although the diode is only just biased above the barrier potential, a large number of carriers are available for conduction, and this causes a high current density.

3.2   Light Emission in Semiconductors

In this section we will examine light generation in semiconductor diodes. In particular we will discuss the rate at which the semiconductor can generate light and comment on the efficiency of certain semiconductor materials. Before we derive the rate equations, it will be useful for us to examine direct and indirect band-gap materials.

3.2.1   Direct and Indirect Band-Gap Materials

As we have already seen, when we apply forward bias to a semiconductor diode, the barrier voltage of a p-n semiconductor junction diode reduces, so allowing electrons and holes to cross the depletion region (Fig. 3.6). The minority carriers, electrons in the p-type and holes in the n-type, recombine by electrons dropping down from the conduction band to the valence band. This recombination results in the electrons losing a certain amount of energy equal to the band-gap energy difference, E_g.

Recombination can occur by two different processes: *indirect transitions* (also known as *non-radiative* recombinations) which produce lattice vibrations, or

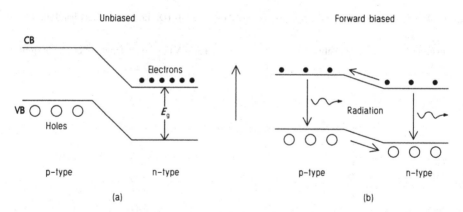

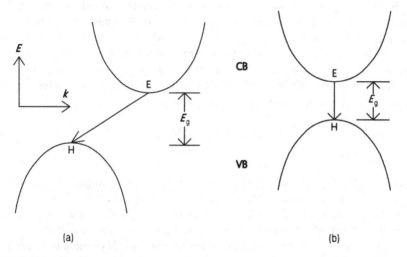

Fig. 3.6 Carrier recombination in a forward biased p-n junction diode

Fig. 3.7 Energy/wave-number diagrams for (**a**) an indirect and (**b**) a direct band-gap semiconductor

phonons, and *direct transitions* (or *radiative* recombinations) which produce *photons* of light. We can see the difference between these by examining the energy/wave number, *E-k*, diagrams of two different semiconductors. (The *E-k* diagrams are plots of electron energy against wave number which we can regard as being proportional to the electron momentum.)

Figure 3.7a shows a simplified *E-k* diagram for an indirect band-gap material, for example, silicon, *Si*, or germanium, *Ge*. As can be seen, the electron and hole momenta are different. So, if an electron drops down from region *E* in the CB to region *H* in the VB, then a change of momentum has to take place, and this results in the emission of a phonon.

Table 3.1 Characteristics of various semiconductor materials (D, direct, I, indirect band-gap)

Semiconductor material	Transition type	Band-gap energy (eV)	Wavelength of emission (μm)
InAs	D	0.36	3.44
PbS	I	0.41	3.02
Ge	I	0.67	1.85
GaSb	D	0.72	1.72
Si	I	1.12	1.11
InP	D	1.35	0.92
GaAs	D	1.42	0.87
CdTe	D	1.56	0.79
GaP	I	2.26	0.55
SiC	I	3.00	0.41

Direct band-gap semiconductors can be made from compounds of elements from groups III and V of the periodic table (so-called *III-V semiconductors*). Gallium arsenide, *GaAs,* is a direct band-gap material, with an *E-k* diagram similar to that shown in Fig. 3.7b. As can be seen, the electron and hole momenta are the same – the regions *E* and *H* are coincident. Thus an electron dropping down from the CB to the VB does so *directly.* Under these circumstances, the energy lost is given up as a photon of light whose free-space wavelength is given by

$$\lambda_o = \frac{hc}{qE_g} = \frac{1244}{E_g} \, (\text{nm}) \tag{3.39}$$

where h is Planck's constant, 6.624×10^{-24} Js, E_g is the band-gap in electronvolts, eV, and q is the electronic charge, 1.6×10^{-19} C. Thus, to be an efficient semiconductor light source, the LED or laser should be made of a direct band-gap material.

Table 3.1 lists the band-gap energy, and transition type, of a range of semiconductors. We can see from this that all the common single element materials have an indirect band-gap and so are never used as light sources. However, the III-V semiconductors have a direct band-gap and so are most often used. For example, a compound of gallium, aluminium and arsenic has a band-gap of between 1.38 and 1.55 eV, resulting in light of wavelength in the region 900–800 nm. The 100 nm spread in wavelength occurs because E_g depends on the ratio of Ga to Al. (We will return to this point presently.)

3.2.2 Rate Equations

The previous section introduced the basic principle of photon generation in semiconductor diodes. However, we have not yet examined the rate of light generation, and this is the subject that will now be considered. To simplify the following analysis, we will initially assume that the electron and hole densities in either side of the junction are constant.

Photon generation occurs by electrons in the CB recombining with holes in the VB. As we saw when we considered the zero biased diode, recombination of carriers can occur on both sides of the junction. If the material is a direct band-gap one, each carrier recombination will, ideally, produce a photon. Thus we can see that the rate of photon generation will depend on several factors: the density of electrons in the CB, n; the density of holes in the VB, p; and absorption by the material. Taking account of these factors, we can write the rate of photon generation, dϕ/dt, as

$$\frac{d\phi}{dt} = anp - b\phi \tag{3.40}$$

where a and b are constants relating to photon generation and absorption, respectively, and ϕ has units of photons/m^3.

Now, each photon consumes one electron-hole pair, *EHP*, and so we can write the rate of decrease in electron density as

$$\frac{dn}{dt} = -anp + b\phi \tag{3.41}$$

Equations (3.40) and (3.41) are non-linear in form. However, we can generate an approximate solution by considering the steady-state solution and then introducing a transient.

In equilibrium, the rate of photon generation equals the rate of change of electron density. So,

$$an_e p_e = b\phi_e \tag{3.42}$$

If we disturb the equilibrium with an injection of electrons, (3.41) becomes

$$\frac{d}{dt}(n_e + \delta n) = -a(n_e + \delta n)(p_e + \delta p) + b(\varnothing_e + \delta\varnothing) \tag{3.43}$$

If we use (3.42), and note that δn equals δp, we get

$$\frac{d}{dt}\delta n = -a\delta n(n_e + p_e) + b\delta\varnothing \tag{3.44}$$

where we have assumed that $\delta n\delta p \approx 0$. The first term in (3.44) is the rate of change of electron density due to photon generation, which we can write as

$$\frac{d}{dt}\delta n = -\frac{\delta n}{\tau_r} \tag{3.45}$$

where τ_r is the *radiative recombination time* defined by

$$\tau_r = \frac{1}{a(n_e + p_e)} \qquad (3.46)$$

Unfortunately, not all EHPs contribute to the light output; some electrons may fall into traps in the crystalline structure so generating phonons rather than photons. Thus we need to introduce *non-radiative recombination*. These two recombination rates will add together, and so we can modify (3.45) to give

$$\frac{d}{dt}\delta n = -\frac{\delta n}{\tau_r} - \frac{\delta n}{\tau_{nr}}$$

or

$$\frac{d}{dt}\delta n = -\frac{\delta n}{\tau_n} \qquad (3.47)$$

where τ_n is the *electron* recombination time given by

$$\frac{1}{\tau_n} = \frac{1}{\tau_r} + \frac{1}{\tau_{nr}} \qquad (3.48)$$

(This τ_n is identical to the recombination time for electrons that we introduced in Sect. 3.1.4.)

The solution to (3.47) is an exponential given by

$$\delta n(t) = \delta n(0)\exp\left(-t/\tau_n\right) \qquad (3.49)$$

We can now write the electron rate equation, (3.44), as

$$\frac{d}{dt}\delta n = -\frac{\delta n}{\tau_n} + b\delta\varnothing \qquad (3.50)$$

By following a similar argument with the photon rate equation, (3.40) can be written as

$$\frac{d\varnothing}{dt} = \frac{\delta n}{\tau_r} - b\delta\varnothing \qquad (3.51)$$

From these two equations, we can see that an increase in photon density will increase the electron density (Eq. 3.50) and this will increase the photon density (Eq. 3.51). Thus (3.50) and (3.51) are intimately linked.

Let us now consider *forward bias operation.* As we saw in Sect. 3.1.4, a current will flow due to EHP recombination when we forward bias the diode. Thus we can modify the electron rate equation to

$$\frac{d}{dt}\delta n = \frac{1}{q}\frac{dJ}{dx} - \frac{\delta n}{\tau_n} + b\delta\varnothing \qquad (3.52)$$

This equation describes the variation in electron density due to carrier injection, EHP recombination and photon absorption. If we consider photon generation due to high injection currents, we can effectively neglect photon absorption. The rate equations then become

$$\frac{d\varnothing}{dt} = \frac{\delta n}{\tau_r} \qquad (3.53)$$

and

$$\frac{d}{dt}\delta n = \frac{1}{q}\frac{dJ}{dx} - \frac{\delta n}{\tau_n} \qquad (3.54)$$

Thus we can see that an increase in current causes an increase in photon density, so producing light. This is the principle behind light-emitting diodes, *LEDs,* which are dealt with in detail later.

We should also note that the injected carriers will modify the radiative recombination time, (3.46), to

$$\tau_r = \frac{1}{a(n_e + p_e + \delta n)} \qquad (3.55)$$

We now have two regions of interest: under high current injection, $\delta n \gg (n_e + p_e)$ and τ_r depends on the injected carrier concentration; under low current injection, $\delta n \ll (n_e + p_e)$ and τ_r is independent of the injected carriers.

We can determine the efficiency of any particular light source by finding the ratio of radiative recombination rate to the total recombination rate. So, the efficiency, η, of a light source is

$$\begin{aligned}\eta &= \frac{\delta n/\delta\tau_r}{\delta n/\delta\tau_n} \\ &= \frac{\tau_n}{\tau_r} \\ &= \frac{\tau_{nr}}{\tau_r + \tau_{nr}}\end{aligned} \qquad (3.56)$$

Thus we can see that, in order to be an efficient light source, the non-radiative recombination lifetime must be far higher than the radiative recombination lifetime. We will return to this point when we consider LEDs and lasers.

Example
A semiconductor light source is made from the p^+n GaAs diode described in the previous example. The voltage across the diode is pulsed from 0.9 volt to 1.2 volts. Determine the optical power generated by the device. (Assume a cross-sectional area of 4×10^{-9} m^2.)

By following a similar analysis to that used in the previous example, we find

$$n(x) - n_p = 8.06 \times 10^{15} \exp\left(-30 \times 10^3 x\right) \text{ m}^{-3}$$

$$p(x) - p_n = 8.06 \times 10^{18} \exp\left(-140 \times 10^3 x\right) \text{ m}^{-3}$$

for a diode bias of 0.9 V and

$$n(x) - n_p = 9.5 \times 10^{20} \exp\left(-30 \times 10^3 x\right) \text{m}^{-3}$$
$$p(x) - p_n = 9.5 \times 10^{23} \exp\left(-140 \times 10^3 x\right) \text{ m}^{-3}$$

for a bias of 1.2 V.

It should be evident from these calculations that the injected carrier density is far larger than the equilibrium carrier density.

Now, the carrier density is a function of distance from the junction. So, we need to carry this variation with x throughout our calculations. As we are operating in the high injection regime, the radiative recombination time is

$$\tau_r = \frac{1}{a \delta n}$$

The constant a is of the order of 10^{-16} m^3/s and so

$$\tau_r = 10.5 \times 10^{-6} \exp\left(30 \times 10^3\, x\right) \text{ for electrons}$$
$$\tau_r = 10.5 \times 10^{-9} \exp\left(140 \times 10^3\, x\right) \text{ for holes}$$

Now, by using (3.53), we find

$$\frac{d\varnothing}{dt} = \frac{\delta n}{\tau_r}$$

$$= \frac{9.5 \times 10^{20} \exp\left(-30 \times 10^3 x\right)}{10.5 \times 10^{-6} \exp\left(-30 \times 10^3 x\right)}$$

$$= 9.05 \times 10^{25} \exp\left(-60 \times 10^3 x\right) \text{m}^3/\text{s for electrons}$$

and

(continued)

$$\frac{d\varnothing}{dt} = 9.05 \times 10^{31} \exp\left(-280 \times 10^3 x\right) \text{ m}^3/\text{s for holes.}$$

In order to find the rate of photon generation per unit area, we need to integrate these flux densities with respect to x. So, for electrons we have

$$\frac{d\varnothing}{dt} = \frac{-9.05 \times 10^{25}}{60 \times 10^3} \int_0^{70\mu m} \exp\left(-60 \times 10^3 x\right)$$
$$= 1.5 \times 10^{21} \text{ m}^2/\text{s}$$

while for holes we have

$$\frac{d\varnothing}{dt} = \frac{-9.05 \times 10^{31}}{280 \times 10^3} \int_0^{330\mu m} \exp\left(-280 \times 10^3 x\right)$$
$$= 3.2 \times 10^{26} \text{ m}^2/\text{s}$$

We also need to take account of the efficiency of the device. As we have already seen

$$\eta = \frac{\tau_n}{\tau_r}$$

and so the efficiency of electron recombination is

$$\eta(x) = \frac{50 \times 10^{-9}}{10.5 \times 10^{-6} \exp\left(30 \times 10^3 x\right)}$$

On integrating this efficiency across the p-type, we get

$$\eta = 1.6 \times 10^{-7} \text{ for the electrons.}$$

By following a similar procedure with hole recombination in the n-type, we get

$$\eta = 3.4 \times 10^{-5} \text{ for the holes.}$$

Hence the total photon generation rate per unit area is

(continued)

$$\frac{d\phi}{dt} = \left(1.5 \times 10^{21} \times 1.6 \times 10^{-7}\right) + \left(3.2 \times 10^{26} \times 3.4 \times 10^{-5}\right)$$
$$= 1.1 \times 10^{22} \ \mathrm{m^2/s}$$

As the area of the device is $4 \times 10^{-9} \ \mathrm{m^2}$, we get a photon generation rate of 4.4×10^{13}. Each photon carries energy of hf Joules, and so the power generated by the diode is 10 μW.

This example has shown that, for a p$^+$n diode, it is the hole recombination in the n-type that is the dominant light-generating mechanism. We have also seen that the efficiency of the device is very low.

In the next section we will consider heterojunction diodes – the most widely used type of semiconductor diode for light generation.

3.3   Heterojunction Semiconductor Light Sources

As we have seen in the previous section, light emission can occur on both sides of the p-n junction. We also saw that the efficiency of the diode was very low. However, if we concentrate the recombining carriers to a small active area, the light output will increase, and we can launch more power into a fibre. We can achieve such confinement by forming a junction between two dissimilar band-gap material – a *heterojunction* – which results in certain carriers experiencing a potential step, so inhibiting them from travelling farther through the lattice. In order to confine both holes and electrons, we must use two heterojunctions, the so-called *double-heterojunction,* or *DH* structure. Although most LEDs and lasers use this structure, we will initially examine a single heterojunction, or *SH,* diode.

Figure 3.8 shows the energy diagram of an SH diode. This particular diode is made of wide band-gap $Ga_{0.8}Al_{0.2}As$ and narrow band-gap GaAs. (The numerical subscripts refer to the proportions of the various elements that make up the alloy.) Such diodes are normally called P-n, or N-p, where the capital letter denotes the material with the higher band-gap. (The most widely used dopants are sulphur, *S,* for n-type and zinc, *Zn,* for p-type.) We can see from the diagram that the potential step for holes, δE_v, is lower than the potential step for electrons, δE_c. This is more obvious when the diode is under forward bias (Fig. 3.8b). So, under forward bias, injected holes travel into the n-type region, but electrons cannot cross into the p-type. Hence there are a great number of holes in the GaAs n-type, and these recombine within a diffusion length of the junction. This area is known as the *active region* and, as the recombination occurs in GaAs, it generates 870 nm wavelength light.

A double heterojunction, *DH,* structure will confine both holes and electrons to a narrow active layer. As Fig. 3.9 shows, the potential steps either side of the active

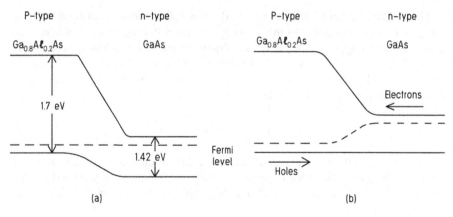

Fig. 3.8 Energy diagram of a heterojunction under (**a**) zero bias and (**b**) forward bias

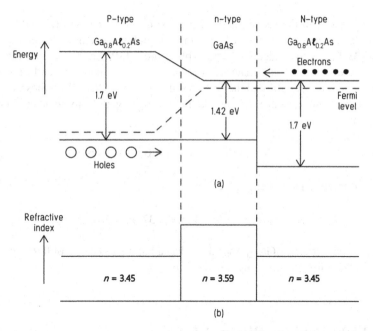

Fig. 3.9 (**a**) Energy diagram and (**b**) refractive index profile for a forward biased P-n-N, double heterojunction diode

region, the GaAs, inhibit carrier movement. Thus, under forward bias, there will be a large number of carriers injected into the active region where they are effectively confined. Carrier recombination occurs in this small active layer so leading to an efficient device. An additional advantage of the DH structure is that the refractive index of the active region is greater than that of the surrounding material. Hence light emission occurs in an optical waveguide, which serves to narrow the output beam.

GaAs emits light at 870 nm; however, the first optical window occurs at 850 nm. The addition of aluminium to the GaAs layer causes the band-gap, and hence the emission wavelength, to change. Hence diodes for the first window are commonly made of an $Al_xGa_{1-x}As$ active layer, surrounded by $Al_yGa_{1-y}As$ with $y > x$. This alloy is a direct band-gap semiconductor for $x < 0.37$. If $0 < x < 0.45$, we can find E_g from the following empirical relationship:

$$E_g = 1.42 + 1.25x + 0.27x^2 \qquad (3.57)$$

As it is the active layer that emits the light, the surrounding material can be an indirect band-gap semiconductor. As an example, a diode with $x = 0.03$ and $y = 0.2$ will emit light of wavelength 852 nm. We can find the refractive index of the material from

$$n = 3.59 - 0.71x \quad \text{for } 0 < x < 0.45 \qquad (3.58)$$

For operation in the second and third transmission windows, 1.3 and 1.55 μm, the diode is usually made of an indium-gallium-arsenide-phosphide alloy, $In_{1-x}Ga_xAs_yP_{1-y}$, surrounded by indium phosphide, InP. To ensure that the active region is a direct band-gap material, x should be lower than 0.47 and, in order to match the active layer alloy to the InP crystal lattice, $y = 2.2x$. With these values of x and y, we can estimate the active region band-gap from another empirical relationship

$$E_g = 1.35 - 1.89x + 1.48x^2 - 0.56x^3 \qquad (3.59)$$

with the refractive index being given by

$$n^2 = 9.6 + 4.52x - 37.62x^2 \qquad (3.60)$$

As an example, $In_{0.74}Ga_{0.26}As_{0.56}P_{0.44}$ has a band-gap energy of 0.95 eV, which results in an emission wavelength of 1.3 μm.

3.4   Light-Emitting Diodes (LEDs)

At present there are two main types of LED used in optical fibre links: the *surface-emitting* LED and the *edge-emitting* LED, or *ELED*. Both devices use a DH structure to constrain the carriers and the light to an active layer. Table 3.2 compares some typical characteristics of the two LED types. From this table we can see that ELEDs are superior to surface emitters in terms of coupled power and maximum modulation frequency. For these reasons, surface emitters are generally used in short-haul, low data-rate links, whereas ELEDs are normally found in medium-haul routes. (Lasers are normally used in long-haul routes.) We should note that LEDs emit light over a

Table 3.2 Comparison of surface- and edge-emitting LED characteristics

LED type surface	Maximum modulation frequency (MHz)	Output power (mW)	Fibre coupled power (mW)
Emitting	60	<4	<0.2
Edge emitting	200	<7	<1.0

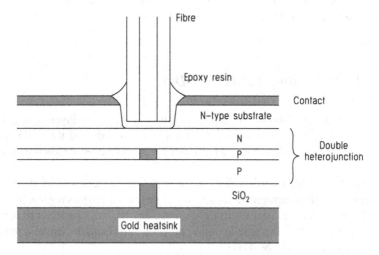

Fig. 3.10 Cross-section through a typical surface-emitting LED

wide area. Thus these devices can usually only couple useful amounts of power into large numerical aperture, MM fibres.

3.4.1 Surface-Emitting LEDs

Figure 3.10 shows the structure of a typical surface-emitting LED. The DH diode is grown on an N-type substrate, at the top of the diode, which has a circular well etched into it. In this particular design, the light produced by the active region travels through the substrate and into a large-core optical fibre held in place by epoxy resin. Some designs dispense with the fibre entirely, preferring to rely on the LED package to guide the light.

At the back of the device is a gold heatsink which, apart from a small circular contact, is insulated from the diode. This heatsink forms one of the contacts, and so all the current flows through the hole in the insulating layer. The current flows through the P-type material and forms a small, circular active region, with a typical current density of 2000 A/cm^2. This results in the production of an intense beam of light.

The refractive index change across the heterojunctions serves to constrain some of the emitted light to the active region. This light is either absorbed, or finally emitted in an area greater than the fibre core. Hence the actual amount of light coupled into the fibre is considerably less than that emitted by the LED. Although a micro-lens placed in the well at the top of the device will increase the coupled power, the efficiency of this arrangement is dependent on the correct truncation of the lens and the fibre alignment. In practice the launched power is two to three times that achieved by an equivalent butt-coupled LED.

3.4.2   Edge-Emitting LEDs (ELEDs)

In order to reduce the losses caused by absorption in the active layer, and make the beam more directional, we can take the light from the edge of the LED. Such a device is known as an *edge-emitting LED,* or *ELED,* and a typical structure is shown in Fig. 3.11.

As can be seen, the narrow stripe on the upper contact defines the shape of the active region. As the heterojunctions act to confine the light to this region, the output is more directional than from a surface-emitting device, and this leads to a greater launch power. A further increase in output power results from the use of a reflective coating on the far end of the diode.

3.4.3   Spectral Characteristics

As we saw in Sect. 3.2, light emission is due to electrons randomly crossing the band-gap, so-called *spontaneous emission* of light. In practice, the conduction and valence bands consist of many different energy levels (Fig. 3.12). It is therefore

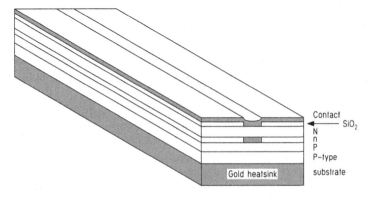

Fig. 3.11   Structure of an edge-emitting, N-n-P, double heterojunction, stripe-contact LED

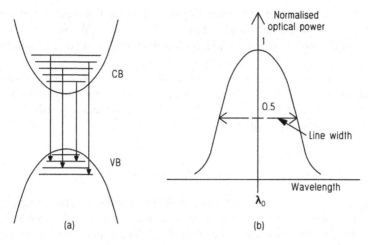

Fig. 3.12 (a) Photon emission from conduction band energy levels and (b) resultant spectral characteristic

possible for recombinations to occur across a wide range of energy differences. The distribution in electron densities peaks at an energy of approximately $E_g + kT/2$ and that of the hole densities at an energy of approximately $E_g - kT/2$. Thus the energy difference has a mean of $E_g + kT$ and a deviation of δE_g which is typically between kT and $2kT$. Although the actual deviation is dependent on the amount of impurity doping, the approximation will suit our purposes.

The spread in recombination energy results in a spread of emitted wavelengths about a nominal peak, as shown in Fig. 3.12b. The half power wavelength spread is known as the *source linewidth* and, as we saw in the previous chapter, a large linewidth will result in considerable material dispersion. However, LEDs can launch a large number of modes into the fibre, and so modal dispersion is usually dominant. In most LEDs, the linewidth is typically 30 nm which translates to a frequency spread of approximately 1.3×10^{13} Hz! Clearly LEDs are not the optical equivalent of an r.f. oscillator; they are, however, useful for simple intensity modulation such as that used in analogue and PCM links.

3.4.4 Modulation Capabilities and Conversion Efficiency

The output power/drive current characteristic of an LED is approximately linear. If we superimpose an a.c. signal on to a d.c. bias level, we can write the output optical power, $p(\omega)$, as

$$p(\omega) = \frac{p(0)}{\sqrt{1 + (\omega\tau)^2}} \tag{3.61}$$

where $p(0)$ is the unmodulated power output and τ is the time constant of the LED and drive circuit. When we considered optical fibre bandwidth, we saw that a 3 dB drop in optical power corresponds to a 6 dB drop in electrical power. Therefore the 3 dB *electrical* bandwidth of the LED is $1/2\pi\tau$ Hz.

With careful design of the drive circuit, the dominant time constant will be that of the LED. This is governed by the recombination time of the carriers in the active region, τ_n. As we have already seen, when both radiative and non-radiative recombinations are present, τ_n is given by

$$\frac{1}{\tau_n} = \frac{1}{\tau_r} + \frac{1}{\tau_{nr}} \tag{3.62}$$

where τ_r and τ_{nr} are the radiative and non-radiative recombination times, respectively. These time constants also give us a measure of the diode conversion efficiency, which we briefly examined in Sect. 3.2.2. The *internal quantum efficiency*, η_{int}, is given by

$$\eta_{int} = \frac{\tau_{nr}}{\tau_{nr} + \tau_r}$$

So in order to produce a fast device, both τ_r and τ_{nr} should be kept low, with the proviso that $\tau_{nr} \gg t_r$, in order to keep the efficiency high.

Let us now return to the electron rate equation, previously given as (3.54):

$$\frac{d}{dt}\delta n = \frac{1}{q}\frac{dJ}{dx} - \frac{\delta n}{\tau_r}$$

If we assume that the semiconductor is lightly doped, and we consider the steady-state condition, that is we apply a current of J which injects an electron density of δn, we get

$$0 = \frac{1}{q}\frac{dJ}{dx} - \frac{\delta n}{\tau_r}$$

and so

$$J \approx \frac{q\delta n d}{\tau_r} \tag{3.64}$$

where d is the distance between the heterojunctions and we have temporarily ignored non-radiative recombination. Now, under high levels of injection, the radiative recombination time, Eq. (3.55), becomes

$$\tau_r = \frac{1}{a\delta n}$$

and so (3.64) becomes

$$J = \frac{qd}{a\tau_r^2}$$

Hence

$$\tau_r \propto \sqrt{\frac{d}{J}} \tag{3.65}$$

From (3.65) we can see that in order to reduce the radiative recombination time, and so produce a more efficient device, we should operate with high current densities.

Let us now take account of non-radiative recombinations. When a heterojunction diode is formed, there is a slight mismatch between the heterojunction crystal lattices. This introduces traps at the interface between the two materials, characterised by the *surface recombination velocity, S.* Thus

$$\tau_{nr} \propto \frac{d}{S} \tag{3.66}$$

As τ_r and τ_{nr} are dependent on d, a smaller d will result in lower time constants. Unfortunately, a reduction in d causes τ_{nr} to fall faster than t_r, and so the modulation speed increases at the expense of the efficiency. However, τ_r is inversely proportional to $\sqrt{J}$, and so we could reduce τ_r by increasing the current density. The problem with this is that a high current density causes difficulties with heatsinking, which tends to impair the device lifetime.

From (3.55) we can see that for high doping levels, $>10^{24}$ m^{-3}, τ_r is inversely proportional to the doping level. So we could reduce τ_r by increasing the doping. Unfortunately, this tends to increase the number of non-radiative recombination centres, and so τ_{nr} will also reduce. Therefore there is a trade-off between the modulation bandwidth and the LED efficiency. Most LEDs operate with high doping levels, and current state-of-the-art devices have a typical internal quantum efficiency of 50 per cent. In spite of this, the external efficiency (a measure of the launch power into a fibre) is typically less than 10 per cent, and so LEDs are generally low-power devices.

3.5 Semiconductor Laser Diodes (SLDs)

Unlike LEDs, which emit light spontaneously, lasers produce light by *stimulated emission.* Stimulated emission occurs when a photon of light impinges on an already excited atom and, instead of being absorbed, the incident photon causes an electron to cross the band-gap, so generating another photon (Fig. 3.13). The stimulated photon has the same frequency and phase as the original, and these two generate more photons as they travel through the lattice. In effect, the lattice amplifies the

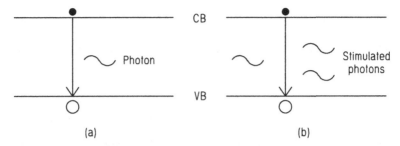

Fig. 3.13 Light generation by (**a**) spontaneous emission and (**b**) stimulated emission

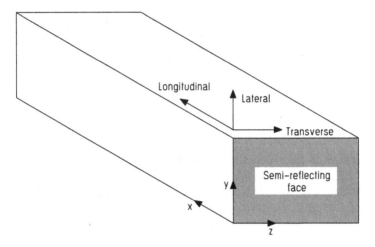

Fig. 3.14 A basic Fabry-Perot cavity

original photon; indeed, the acronym laser stands for light amplification by the stimulated emission of radiation. As the generated photons are all in phase, the light output is coherent and has a narrow linewidth.

Before stimulated emission can occur, the CB must contain a large number of electrons and the VB a large number of holes. This is a *quasi-stable* state known as a *population inversion*. It results from the injection of a large number of carriers into a heavily doped, ELED active layer. If a population inversion is present then, by virtue of the light confinement from the heterojunctions, some stimulated emission occurs. However, in order to ensure that it is the dominant light-generating process, we must provide some additional optical confinement.

In a laser diode, the extra confinement results from cleaving the end faces so that they form partial reflectors, or *facets*. The resulting structure, known as a *Fabry-Perot etalon*, is shown in schematic form in Fig. 3.14. The facets reflect some of the spontaneously emitted light back into the active region, where it causes stimulated emission and hence gain. So, provided the optical gain in the cavity exceeds the losses, stimulated emission will be dominant.

3.5.1 Stimulated Emission

Laser diodes and LEDs differ in several ways: a laser diode requires the application of a constant current to maintain stimulated emission; the output beam is more directional; and the response time is faster. In this section we will examine the simple stripe contact laser, which is similar in construction to a stripe contact ELED. We will deal with other SLD structures in 3.5.4. We begin our study by examining the optical gain that stimulated emission produces. We will then go on to study the spectral characteristics of SLDs. It should be evident that, because light emission occurs in a rectangular cavity, propagation can occur along all three axes; *longitudinal, transverse,* and *lateral* propagation. Let us initially consider a longitudinal TE wave given as $E(x, t)$. If we neglect the effects of the cavity side walls, and assume that the cavity confines all of the E field, then we can write $E(x, t)$ as

$$E(x,t) = |E| \exp\left(-\alpha x/2\right) \exp j(\omega t\text{-}\beta_1 x) \qquad (3.67)$$

where α is the attenuation of the optical *power* per unit length (hence the factor 1/2) and β_1 is the phase constant in the active region. So, the field just to the right of the mirror at $x = 0$ is

$$E(0,t) = \exp\left(0\right) \exp j\omega t \qquad (3.68)$$

Now, when the field undergoes a round trip of distance $2\,L$, it is reflected off both mirrors and amplified by stimulated emission. Thus after one round trip, the travelling field, E_r, is

$$E_r(0,t) = \sqrt{R_1 R_2}|E| \exp\left[(g - \alpha)L\right] \exp j(\omega t - 2\beta_1 L) \qquad (3.69)$$

where R_1 and R_2 are the *reflectivity* of the mirrors at $x = 0$ and $x = L$, respectively, and g is the *power* gain per unit length. (The reflectivity is defined as the ratio of the reflected to the incident power, hence the presence of the square root.) For amplification to occur, the magnitude of the reflected wave must be greater than that of the original wave, that is

$$\sqrt{R_1 R_2}|E| \exp\left[(g - \alpha)L\right] \geq |E|$$

Therefore the optical gain for lasing is given by

$$g \geq \alpha + \frac{1}{2L} \ln \frac{1}{R_1 R_2} \qquad (3.70)$$

As we have already noted, it is the current density in the active region, J, that produces the population inversion and hence the cavity gain. In order to determine the relationship between the gain and current density, we must consider the rate equations for a SLD. A full analysis of stimulated emission in laser diodes would involve us in the black-body radiation law and quantum mechanics in general. This is beyond the scope of this text; instead we will quote necessary results where needed.

As we saw when we considered light emission in semiconductors, electrons dropping down from the CB to the VB generate photons. There are two ways in which this can occur: spontaneous emission and stimulated emission. If we have a VB electron density of n_2, we can express the rate of electron density decay due to *spontaneous* recombination as

$$\frac{dn}{dt}_{\text{spon}} = \frac{n_2}{\tau_{\text{sp}}} = A_{21}n_2 \tag{3.71}$$

where A_{21} is a constant with units of s^{-1}.

Now, the semiconductor will absorb some of the light generated by spontaneous emission. As we have seen when we considered light emission in an LED, spontaneous emission generates light with a spread of wavelength. Thus we have light of frequency f_o and half-power spread δf propagating through the SLD. One way of finding δf is to excite the SLD material with light of varying wavelength and plot the variation of absorption with frequency. The result is a Lorentzian curve, $g(f)$, given by

$$g(f) = \frac{\delta f}{2\pi\left[(f-f_o)^2 + (\delta f/2)^2\right]} \tag{3.72}$$

where we have used the following normalisation

$$\int_{-\infty}^{\infty} g(f)df = 1 \tag{3.73}$$

Thus the units of $g(f)$ are 1/Hz or seconds.

Some of the light generated by spontaneous emission will be absorbed by the SLD material. This will have the effect of increasing the electron density in the VB at a rate

$$\frac{dn}{dt}_{\text{abs}} = B_{12}\emptyset hfn_1g(f) \tag{3.74}$$

where B_{12} is another constant with units of m^3/Js^2, ϕ is the optical flux density in photons/m^3, and n_1 is the electron density in the CB. (We have included the factor hf because we are considering photon density.)

Electrons will also be lost due to stimulated recombination. So

$$\left.\frac{dn}{dt}\right|_{stim} = B_{21}\varnothing hfn_2 g(f) \tag{3.75}$$

The constants A_{21}, B_{12}, and B_{21} are known as the Einstein A and B coefficients. (Albert Einstein was the first person to suggest the possibility of stimulated emission.) If the semiconductor is in equilibrium, we can write

$$\left.\frac{dn}{dt}\right|_{abs} = \left.\frac{dn}{dt}\right|_{spon} + \left.\frac{dn}{dt}\right|_{stim}$$

or

$$B_{12}\varnothing hfn_1 g(f) = A_{21}n_2 + B_{21}\varnothing hfn_2 g(f)$$

Quantum mechanics predicts $B_{12} = B_{21} = B$ and that

$$\frac{A}{B} = \frac{8\pi hf^3}{v_g^3} \tag{3.76}$$

where f is the frequency of the photons and v_g is the group velocity of the light in the semiconductor material. Now, we can express the net decrease in electron density due to stimulated emission, from (3.74) and (3.75), as

$$\begin{aligned}\left.\frac{dn}{dt}\right|_{loss} &= (B_{21}n_2 - B_{12}n_1)\varnothing hfg(f)\\ &= (n_2 - n_1)B\varnothing hfg(f)\end{aligned} \tag{3.77}$$

This decrease in VB electron density causes a corresponding increase in the stimulated photon density and so

$$\frac{d\varnothing}{dt} = (n_2 - n_1)B\varnothing hfg(f)$$

These photons are emitted at a velocity of v_g along the x-axis and so the rate of photon emission *per unit area* is

$$\frac{d\varnothing}{dt} = (n_2 - n_1)B\varnothing hfg(f)dx$$

Thus the power emitted per unit area is

$$dP = (n_2 - n_1)B\varnothing hfg(f)dx\, hf \tag{3.78}$$

Emitted photons pass through this unit area at a velocity of v_g and so we can write the power per unit area as

$$P = \emptyset hf\, v_g$$

or

$$\emptyset = \frac{P}{hf\, v_g} \tag{3.79}$$

We can substitute this into (3.78) to give

$$dP = (n_2 - n_1)B\frac{P}{hf\, v_g}hfg(f)dx\, hf$$

$$= (n_2 - n_1)\frac{BP}{v_g}g(f)\, hfdx$$

Therefore

$$\frac{dP}{P} = (n_2 - n_1)\frac{B}{v_g}g(f)\, hfdx \tag{3.80}$$

The solution to (3.80) is an exponential given by

$$P(x) = P(0)\exp(gx) \tag{3.81}$$

where

$$g = (n_2 - n_1)\frac{B}{v_g}g(f)\, hf \tag{3.82}$$

We can eliminate B from this equation by using (3.76) and by noting that $A = 1/\tau_{sp}$, where τ_{sp} is the carrier lifetime for spontaneous emission. Thus

$$g = (n_2 - n_1)\frac{v_g^2}{8\pi f^2 \tau_{sp}}g(f)$$

or

$$g = (n_2 - n_1)\frac{\lambda_o^2}{8\pi \varepsilon_r \tau_{sp}}g(f) \tag{3.82}$$

where ε_r is the relative permittivity of the active region. From (3.82) we can see that n_2 must be greater than n_1 for stimulated emission to occur. This is the quasi-stable state known as a *population inversion*. We can produce a population inversion in SLDs by injecting a large number of electrons into the active region of a double heterojunction diode.

Example

A semiconductor laser diode has a GaAs active region and a population inversion of 2.5×10^{24} m^{-3}. Determine the optical gain under these conditions. (Take $\varepsilon_r = 13.1$, $\tau_{sp} = 4$ ns and assume a half-power linewidth of 10 nm for the gain function.)

We can use Eq. (3.82) to express the gain as

$$g = (n_2 - n_1)\frac{\lambda_o^2}{8\pi\varepsilon_r\tau_{sp}}g(f)$$

Now, $g(f)$ is given by Eq. (3.72)

$$g(f) = \frac{\delta f}{2\pi\left[(f - f_o)^2 + (\delta f/2)^2\right]}$$

A linewidth of 10 nm about a centre wavelength of 870 nm gives a frequency spread of 1.1×10^{12} Hz. Thus the gain function *at the centre frequency* is

$$g(f) = \frac{\delta f}{2\pi\left[0 + (\delta f/2)^2\right]}$$
$$= 5.8 \times 10^{-13}\,\text{s}$$

Thus, the optical gain is

$$g = 2.5 \times 10^{24}\frac{(870 \times 10^{-9})^2}{8\pi \times 13 \times 4 \times 10^{-9}} \times 5.8 \times 10^{-13}$$
$$= 8.4 \times 10^{5}\,\text{m}^{-1}$$
$$= 8.4 \times 10^{3}\,\text{cm}^{-1}$$

We should note at this stage that the gain is effectively clamped at a value given by (3.70). Thus, even if we increase the population inversion, the gain cannot increase past the limit given by (3.70).

It should be evident that we must bias the SLD at a certain current to maintain a population inversion. Below this *threshold current,* the SLD will emit light spontaneously as there will not be enough current to generate a

(continued)

population inversion. In order to find the threshold current density, J_{th}, we must study the rate equations for a SLD.

In Sect. 3.2.2 we derived the rate equations for an LED, Eqs. (3.40) and (3.41). As we are considering stimulated emission, we can write the SLD rate equations as

$$\frac{dn}{dt} = \frac{1}{q}\frac{J}{d} - \frac{n_2}{\tau_{sp}} - C(n_2 - n_1)\phi \tag{3.83}$$

and

$$\frac{d\phi}{dt} = C(n_2 - n_1)\phi + \frac{Dn_2}{\tau_r} - \frac{\phi}{\tau_{ph}} \tag{3.84}$$

where C is a constant of proportionality for stimulated emission and τ_{ph} is the stimulated photon lifetime in the active region. We can justify these equations by simple book-keeping. The first term in (3.83) is the injected carrier density; the second term is the number of carriers lost due to recombination; and the third term is the total loss due to stimulated emission and absorption. As regards (3.84) the first term is the total increase in light due to stimulated emission and absorption; the second term is the fraction of spontaneous emission coupled into a laser mode; and the third term is the loss due to photons being emitted by the cavity. (Although the constant D in (3.84) is typically very low ($\approx 10^{-3}$), its presence helps to explain operation below threshold.)

Rather than examine the dynamic behaviour of a SLD, we will initially study the steady-state rate equations. Thus, dn/dt and $d\phi/dt$ are zero, and we can write

$$0 = \frac{1}{q}\frac{J}{d} - \frac{n_2}{\tau_{sp}} - C(n_2 - n_1)\phi \tag{3.85}$$

and

$$0 = C(n_2 - n_1)\phi + \frac{Dn_2}{\tau_r} - \frac{\phi}{\tau_{ph}} \tag{3.86}$$

We can combine these equations to give

$$\frac{\phi}{\tau_{ph}} = \frac{Dn_2}{\tau_r} + \left[\frac{1}{q}\frac{J}{d} - \frac{n_2}{\tau_{sp}}\right] \tag{3.87}$$

The first term in (3.87) is the spontaneous emission term, while the term in the brackets relates to stimulated emission.

(continued)

Now, with a laser diode there are three regions of interest: operation *below* threshold, operation *at* threshold and operation *above* threshold. If we consider operation *below threshold,* the stimulated emission term is zero and so

$$\frac{1}{q}\frac{J}{d} = \frac{n_2}{\tau_{sp}}$$

which implies

$$n_2 = \frac{\tau_{sp}}{q}\frac{J}{d}$$

We should also note that zero stimulated emission implies from (3.87)

$$\frac{\phi}{\tau_{ph}} = \frac{Dn_2}{\tau_r}$$

$$= \frac{D\,\tau_{sp}}{\tau_r}\frac{J}{q}\frac{J}{d}$$

$$= \frac{D\,\tau_{sp}}{\tau_r}\frac{I}{q}\frac{I}{\text{volume}}$$

Thus we can write the output optical power as

$$P = \frac{D}{\tau_r}\frac{\tau_{sp}}{q}Ihf \tag{3.88}$$

Equation (3.88) shows that if we operate a SLD below threshold, the power output is directly proportional to the applied current, that is, *it is operating as an LED.*

Let us now turn our attention to operation *at threshold.* If we increase the drive current, the light output will increase until there is sufficient spontaneously emitted light to cause stimulated emission. At this stage, we can neglect spontaneous emission, and so (3.86) becomes

$$0 = C(n_{th} - n_1)\phi - \frac{\phi}{\tau_{ph}}$$

and so

$$n_{th} = \frac{1}{C\tau_{ph}} + n_1$$

where n_{th} is the threshold electron density. So, provided we know the photon lifetime, we can find n_{th}. To find τ_{ph} we note that C is given by

(continued)

$$C = B \, hf \, g(f)$$

and, from our previous analysis, the optical gain at threshold is (Eq. 3.81)

$$g = (n_{th} - n_1) \frac{Bhf}{v_g} g(f)$$

and so

$$
\begin{aligned}
g &= (n_{th} - n_1) \frac{C}{v_g} \\
&= \left(\frac{1}{C\tau_{ph}} + n_1 - n_1 \right) \frac{C}{v_g} \\
&= \frac{1}{\tau_{ph} v_g}
\end{aligned}
\tag{3.90}
$$

The optical gain at threshold is also given by (3.70) as

$$g \geq \alpha + \frac{1}{2L} \ln \frac{1}{R_1 R_2}$$

and so we can find τ_{ph} from

$$\frac{1}{\tau_{ph}} = v_g \left[\alpha + \frac{1}{2L} \ln \frac{1}{R_1 R_2} \right] \tag{3.91}$$

Thus we can see that the photon lifetime only depends on the physical parameters of the SLD and not on the level of injection. As we will see in the next section, the photon lifetime also sets a limit on the maximum rate of modulation.

We can now find the threshold current density. By substituting (3.91) into (3.89), we get

$$n_{th} = \frac{v_g}{Bhfg(f)} \left[\alpha + \frac{1}{2L} \ln \left(\frac{1}{R_1 R_2} \right) \right] + n_1 \tag{3.92}$$

The bias current supplies these carriers and so

$$J_{th} = \frac{q d n_{th}}{\tau_{sp}} \tag{3.93}$$

If we assume that $n_{th} \gg n_1$, we can write

(continued)

$$J_{th} = \frac{qd}{g(f)} \frac{8\pi\varepsilon_r}{\lambda_o{}^2} \left[\alpha + \frac{1}{2L} \ln\left(\frac{1}{R_1 R_2}\right)\right]$$

and so J_{th} at the nominal wavelength of emission is

$$J_{th} = qd\frac{\pi\delta f}{2} \frac{8\pi\varepsilon_r}{\lambda_o{}^2} \left[\alpha + \frac{1}{2L} \ln\left(\frac{1}{R_1 R_2}\right)\right] \qquad (3.94)$$

We can see from this equation that J_{th} is directly proportional to the width of the active region and the linewidth of the ELED that makes up the SLD. Although (3.94) is reasonably accurate at low temperatures, under normal operating conditions, we can approximate J_{th} by

$$J_{th}(T) = 2.5 J_{th} \exp\left(T/120\right) \qquad (3.95)$$

Example
A 300-μm-long GaAs SLD has a loss per cm of 10 and facet reflectivity of $R_1 = R_2 = 0.6$. Determine the gain required before lasing occurs and the threshold current at this point. Take the width of the active layer to be 5 μm. Assume an operating wavelength of 870 nm, a spontaneous emission linewidth of 10 nm, and $\varepsilon_r = 13.1$.
We can find the gain required for lasing by using (3.70). Thus

$$g = \alpha + \frac{1}{2L} \ln\left(\frac{1}{R_1 R_2}\right)$$
$$= 27 \text{ cm}^{-1}$$

We can now use (3.94) to give the threshold current density as

$$J_{th} = qd\frac{\pi\delta f}{2} \frac{8\pi\varepsilon_r}{\lambda_o{}^2} g$$
$$= 5.9 \times 10^6 \text{ A/m}^2$$
$$= 0.59 \text{ kA/cm}^2 \text{ at low temperatures}$$

If we operate the diode at 300 K, we can use (3.95) to give

(continued)

$$J_{th}(T) = 2.5 J_{th} \exp(T/120)$$
$$= 2.5 \times 0.59 \times 10^3 \times \exp(300/120)$$
$$= 18 \text{ kA/cm}^2 \text{ at 300 Kelvin}$$

Thus

$$I_{th} = 270 \text{ mA}$$

Although these values are typical for gain-guided stripe contact lasers, the actual threshold current is likely to be slightly higher than predicted. This is because we have been assuming that the active region confines all of the generated light. In practice, some light will escape the active region so increasing the threshold requirement.

If we now operate the laser *above threshold,* there is negligible spontaneous emission, and so (3.87) becomes

$$\frac{\phi}{\tau_{ph}} = \left[\frac{1}{q}\frac{J}{d} - \frac{n_2}{\tau_{sp}} \right] \quad (3.96)$$

Now, as the laser is operating above threshold, the gain is effectively clamped at the value given by (3.70) and so any increase in carrier density will not increase the gain – it will, however, increase the light output. Thus we can see that n_2/τ_{sp} is held at its threshold value. Under these conditions, we can write (3.96) as

$$\frac{\phi}{\tau_{ph}} = \left[\frac{1}{q}\frac{J}{d} - \frac{n_{th}}{\tau_{sp}} \right]$$
$$= \frac{1}{q}\left[\frac{J - J_{th}}{d} \right] \quad (3.97)$$

and so the output optical power is

$$P = \frac{I - I_{th}}{q} hf \quad (3.98)$$

Thus we can see that, for the SLD operating above threshold, the output power is directly proportional to the amount of bias current above threshold. In practice, SLDs operating above threshold do not exhibit a strictly linear relationship between light output and bias current. This is because of *mode-hopping* which we will deal with in the next section.

Figure 3.15 shows the measured variation of output power with diode current for a typical 850 nm SLD. This figure clearly shows the threshold point above which the

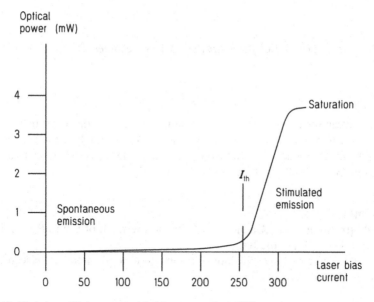

Fig. 3.15 Variation of light output with drive current for a SLD

SLD operates as a laser rather than an ELED. We should also note that the output power saturates at high currents. This is because high currents cause heating of the diode, and this reduces the conversion efficiency.

3.5.2 Spectral Characteristics

In common with the planar optical waveguide, only light waves of certain wavelengths can propagate in the cavity. The condition for successful propagation is that the reflected and original waves must be in phase. At the start of the last section, we found that the field just to the right of the mirror at $x = 0$ is given by Eq. (3.68):

$$E(0, t) = \exp(0) \exp j\omega t$$

We also found that the field travelling back down the cavity, Eq. (3.69), is

$$E_r(0, t) = \sqrt{R_1 R_2} |E| \exp[(g - \alpha)L] \exp j(\omega t - 2\beta_1 L)$$

In order for the wave to propagate successfully, the phase of the two waves must be the same at $x = 0$, that is

$$\exp j(-2\beta_1 L) = 1 \qquad (3.99)$$

Therefore

$$2\beta_1 L = 2\pi N \qquad (3.100)$$

where N is an integer. Since $\beta_1 = 2\pi n_1/\lambda_o$, (3.100) becomes

$$\lambda_o = \frac{2n_1}{N} L \qquad (3.101)$$

Thus we can see that the laser will only amplify wavelengths that satisfy (3.101). Each wavelength is known as a *longitudinal mode,* or simply a mode (not to be confused with the modes in an optical fibre.) The modes cause a line spectrum, and solution of (3.101) will yield the mode spacing.

Example

A 600-μm-long A_l$_{0.03}$Ga$_{0.97}$As SLD has a linewidth of 5 nm. Determine the number of laser modes.

As the active region of the laser is a compound semiconductor, we can use Eqs. (3.57) and (3.58) to give

$$E_g = 1.46\,\text{eV resulting in}\,\lambda_o = 853\,\text{nm}$$

and

$$n = 3.57$$

With these figures, the nominal mode number (from Eq. 3.101) is 5022. The next mode corresponds to $N = 5023$, resulting in a mode spacing of 0.17 nm. Thus, with a linewidth of 5 nm, we have approximately 30 different laser modes of varying wavelength.

The spectral emission of a laser is highly dependent on the bias current. Below threshold, spontaneous emission predominates and so the linewidth is similar to that of an LED. However, if we operate above threshold, we find that the linewidth reduces. This reduction occurs because the cavity exponentially amplifies the first mode to reach threshold, at the expense of all other modes. To see this, let us return to the steady-state solution for the photon density, Eq. (3.86):

$0 = C(n_2 - n_1)\phi + \frac{Dn_2}{\tau_r} - \frac{\phi}{\tau_{ph}}$ We can rearrange this equation to give the photon concentration as

$$\varnothing = \frac{Dn_2}{\tau_r}\left[\frac{1}{\tau_{ph}} - C(n_2 - n_1)\right]^{-1}$$

The term outside the brackets is the amount of spontaneous emission coupled into a laser mode. Thus we can interpret this equation as an amplification factor, G, given by

(continued)

$$G = \left[\frac{1}{\tau_{ph}} - C(n_2 - n_1)\right]^{-1} \tag{3.102}$$

acting on the spontaneous emission of an ELED. As the gain function has a Lorentzian distribution, we can express (3.102) as

$$G = \left[\frac{1}{\tau_{ph}} - \left\{C(n_2 - n_1) + b(\omega - \omega_o)^2\right\}\right]^{-1} \tag{3.103}$$

When we operate a SLD below threshold, the term in { } is small, and so the cavity amplifies all the propagating modes to the same extent. As we increase the diode current, the amplification increases, but the mode whose wavelength is closest to the nominal operating wavelength is amplified the most. This effect is shown in Fig. 3.16. Thus we can see that when we operate a SLD above threshold, the linewidth is considerably less than that of an ELED.

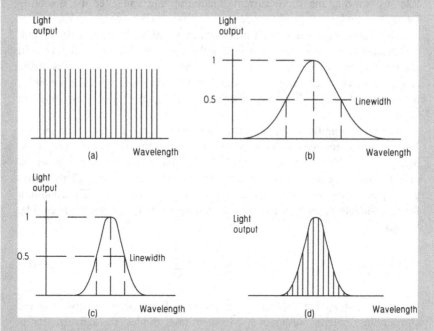

Fig. 3.16 (a) Allowable modes in a SLD; (b) gain profile of a SLD operating below threshold; (c) gain profile of a SLD operating above threshold; and (d) resultant emission spectrum

(continued)

In practice, modes close to the fundamental also undergo significant amplification, and so the output consists of a range of modes following a *gain profile*. We can approximate this profile to the Gaussian distribution

$$g(\omega) = g(\omega_o) \exp\left(\frac{-(\omega - \omega_o)^2}{2\sigma^2}\right) \tag{3.104}$$

where σ is the linewidth of the laser output. This result, together with the line spectrum, causes the emission spectrum shown in Fig. 3.16d. The linewidth of typical stripe contact SLDs can vary from 2 to 5 nm although other structures, examined in 3.5.4, have linewidths measured in kHz.

If we operate the laser at currents significantly higher than threshold, the gain profile may shift slightly so that one of the modes close to the nominal wavelength becomes dominant. This effect is known as *mode-hopping* and it is responsible for kinks in the power/current characteristic. If we modulate the laser by varying the drive current, mode-hopping can alter the operating frequency, and so dynamic mode-hopping is also known as *chirp*. Mode-hopping can cause problems in high-data-rate optical fibre links. If the link is operating at a zero dispersion wavelength, any chirp on the optical pulse will change the operating wavelength so causing pulse dispersion. Thus stripe contact lasers are not commonly found in high-data-rate optical fibre links. Instead we must use alternative laser structures, such as those considered in Sect. 3.5.4.

As well as longitudinal modes, there are also *transverse* and *lateral modes*. These tend to produce an output beam which is highly divergent, resulting in inefficient launching into an optical fibre. The ideal situation is one in which only the fundamental transverse and lateral modes are present. (This would give a parallel beam of light of very small cross-sectional area.) The condition for a single lateral mode is identical to that for a planar dielectric waveguide, and so

$$d < \frac{\lambda_o}{2\sqrt{n_1^2 - n_2^2}} \tag{3.105}$$

where n_1 and n_2 are the refractive indices of the active region and the surrounding material, respectively. In most laser diodes, the active region is typically less than 1 μm thick and (3.105) is usually satisfied. Unfortunately single transverse mode operation is more difficult to achieve. This is because the width of the active region is set by the current density profile in the active layer, which can be difficult to control in the stripe contact lasers we are considering.

3.5.3 Modulation Capabilities

As we have seen, stimulated emission only occurs if a population inversion is present in the active region. It takes a significant length of time for the SLD current to set up a population inversion, and so SLDs are generally biased above threshold using a constant current source.

Let us initially consider digital modulation of the SLD. If we bias the SLD at threshold, and we increase the drive current by a small fraction, δI, we can write

$$n = n_{th} + \delta n \quad \text{and} \quad \phi = \phi_s + \delta\phi$$

where ϕ_s is the steady-state photon density. We can substitute these densities into the rate equations, (3.83) and (3.84) to give

$$\frac{d}{dt}\delta n = -C\delta n\phi_s - \frac{\delta n}{\tau_{sp}} - Cn_{th}\delta\phi \tag{3.106}$$

and

$$\frac{d}{dt}\delta\phi = C\delta n\phi_s \tag{3.107}$$

In deriving (3.106) and (3.107), we have assumed that spontaneous emission is negligible; we have neglected the $\delta n\delta\phi$ term; and we have made use of Eqs. (3.89), (3.93) and (3.97). For reasons of clarity, these equations are reproduced here:

$$n_{th} \approx \frac{1}{C\tau_{ph}}$$

$$J_{th} = \frac{qdn_{th}}{\tau_{sp}}$$

$$\frac{\phi_s}{\tau_{ph}} = \frac{1}{q}\left[\frac{J - J_{th}}{d}\right]$$

We can combine (3.106) and (3.107) by differentiating (3.106) with respect to time and substituting for $d\delta\phi/dt$ from Eq. (3.107). Thus we can write

$$\frac{d^2\delta n}{dt^2} + \left[C\phi_s + \frac{1}{\tau_{sp}}\right]\frac{d}{dt}\delta n + C^2 n_{th}\phi_s\delta n = 0$$

We can rewrite this equation as

$$\frac{d^2 \delta n}{dt^2} + 2\sigma \frac{d}{dt} \delta n + \omega_o{}^2 \delta n = 0 \tag{3.108}$$

where

$$2\sigma = C\phi_s + 1/\tau_{sp}$$
$$= \frac{\phi_s}{n_{th}\tau_{ph}} + \frac{1}{\tau_{sp}} \tag{3.109}$$

and

$$\omega_o{}^2 = C^2 n_{th} \phi_s$$
$$= \frac{\phi_s}{n_{th}\tau_{ph}{}^2} \tag{3.110}$$

Equation (3.108) is a standard differential equation whose solution describes a damped oscillation, thus

$$\delta n = A \sin(\omega t) \exp(-\sigma t)$$

where $\omega^2 = \omega_o{}^2 - \sigma^2$ and A is a constant of integration. We can find A by using the initial conditions $n = n_{th}$ and $\phi = 0$ at $t = 0$. Thus we find

$$A = \frac{\omega}{C}$$

and so

$$\delta n = \frac{\omega}{C} \sin(\omega t) \exp(-\sigma t) \tag{3.111}$$

we can substitute (3.111) into (3.107) to give

$$\delta \phi = -\phi_s \cos(\omega t) \exp(-\sigma t) \tag{3.112}$$

Now, the damping coefficient, σ, is usually small with respect to ω_o, and so we get $\omega \approx \omega_o$ and $A = \sqrt{n_{th}\varnothing_s}$. With this approximation we can write

$$\delta n = \sqrt{n_{th}\varnothing_s} \sin(\omega_o t) \exp(-\sigma t) \tag{3.113}$$

and

Fig. 3.17 Output pulse
from an AlGaAs laser diode

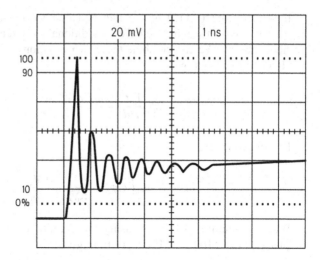

$$\delta\phi = -\phi_s \cos(\omega_o t) \exp(-\sigma t) \tag{3.114}$$

So, under digital modulation, the optical pulse suffers from ringing at an angular frequency given by (3.110). Figure 3.17 shows the actual output pulse of an AlGaAs laser diode operating at 850 nm. As can be seen, the figure clearly shows the damped response predicted by (3.114).

Let us now turn our attention to analogue modulation of the laser diode. Once again we assume that the diode is biased sufficiently above threshold so that the current never falls below I_{th}. If we modulate the diode current at an angular frequency of ω_m, we can write

$$J = J_o + J' \exp(j\omega_m t)$$

We can also express the carrier and photon densities as

$$n = n_o + n' \exp(j\omega_m t) \quad \text{and} \quad \varnothing = \varnothing_o + \varnothing' \exp(j\omega_m t)$$

We can now substitute these values into the rate equations, (3.83) and (3.84), to give

$$j\omega_m n' = \frac{1}{q}\frac{J'}{d} - \frac{n'}{\tau_{sp}} - \frac{\phi'}{\tau_{ph}} - Cn'\phi_o \tag{3.115}$$

and

$$j\omega_m \phi' = Cn'\phi_o \tag{3.116}$$

If we compare (3.115) to the rate equation we derived for a step function, Eq. (3.106), we can see that we have an additional term due to the sinusoidal change in diode current. We can combine these two equations by substituting for n' from (3.116) into (3.115) to give

$$-\omega_m^2 \phi' = C\phi_o \frac{1}{q} \frac{J'}{d} - \frac{j\omega_m \phi'}{\tau_{sp}} - \frac{C\phi_o \phi'}{\tau_{ph}} - C\phi_o j\omega_m \phi'$$

By collecting terms, we can write

$$\phi' \left[-\omega_m^2 + j\omega_m \left(C\phi_o + \frac{1}{\tau_{sp}} \right) + \frac{C\phi_o}{\tau_{ph}} \right] = C\phi_o \frac{1}{q} \frac{J'}{d} \qquad (3.117)$$

We can simplify (3.117) by using the definitions of damping factor, natural frequency, Eq. (3.109) and Eq. (3.110), to give

$$\phi' \left[-\omega_m^2 + j\omega_m 2\sigma + \omega_o^2 \right] = C\phi_o \frac{1}{q} \frac{J'}{d} \qquad (3.118)$$

where we have also made use of $n_{th} = 1/(C\tau_{ph})$. The laser emits photons at a rate of $1/\tau_{ph}$ and so we can write the modulated optical power, P', as

$$P' = \frac{hcI'}{q\lambda_o} \frac{\omega_o^2}{(-\omega_m^2 + j\omega_m 2\sigma + \omega_o^2)} \qquad (3.119)$$

where we have again made use of $n_{th} = 1/(C\tau_{ph})$. We can define the change in optical power under d.c. modulation as

$$P'(0) = \frac{hcI'}{q\lambda_o}$$

and so (3.119) becomes

$$P' = P'(0) \frac{\omega_o^2}{(-\omega_m^2 + j\omega_m 2\sigma + \omega_o^2)}$$

Thus we can write the modulation depth, $m(\omega)$, as

$$m(\omega) = \frac{P'}{P'(0)} = \frac{\omega_o^2}{(-\omega_m^2 + j\omega_m 2\sigma + \omega_o^2)} \qquad (3.120)$$

3.5.4 SLD Structures

As SLDs are normally used in long-haul, high-data-rate routes, which use SM fibres, it is generally desirable to minimise the linewidth (to minimise material dispersion) and operate with a single lateral mode to launch power into SM fibre. It is also important to reduce the threshold current, as this will produce a more efficient device.

At present, the most common SLD structures are the buried heterojunction, *BH*, laser; the distributed feedback, *DFB*, laser; the distributed Bragg reflector, *DBR*, laser; and the vertical cavity surface-emitting laser, *VCSEL*. Apart from the VCSEL, these have evolved from the stripe contact laser we have been considering the aim being to reduce the threshold current and operate in single mode.

The most obvious way of reducing I_{th} is to reduce the active region cross-sectional area. As this is set by the area of the stripe contact, we could reduce I_{th} by reducing the cavity length. Unfortunately this causes the gain required for threshold to increase (see Eq. 3.94) so causing J_{th} to increase. As a high current density causes heatsinking problems, the cavity length is usually limited to typically 150 μm, and so we must reduce the contact width to reduce I_{th}.

To a certain extent the width of the active region is set by the width of the contact. In practice, I_{th} fails to fall in proportion to the contact stripe width, if it is less than about 6 μm. This is because the injected current tends to diffuse outwards as it travels through the laser. Ultimately, we get an active region that is independent of the contact width. So the threshold current of stripe contact lasers is usually no less than 120 mA.

In order to reduce I_{th} further, and operate with a single lateral mode, we must use a different structure. In a *buried heterostructure, BH,* laser, the diode current is constrained to flow in a well-defined active region, as shown in Fig. 3.18. The heterojunctions in either side of the active region provide carrier confinement, and so the width of this region can be made very small, typically 2 μm or less. The heterojunctions will also produce a narrow optical waveguide, and so single lateral mode operation is often achievable. The threshold current of these devices is typically lower than 30 mA.

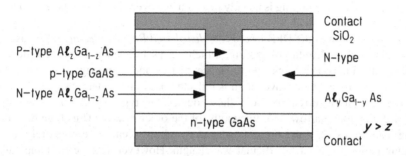

Fig. 3.18 Cross-section through a buried heterojunction, semiconductor laser diode

Fig. 3.19 Cross-section through a distributed feedback semiconductor laser diode

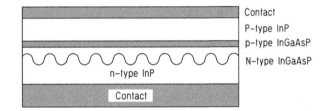

Contact
P–type InP
p–type InGaAsP
N–type InGaAsP
n–type InP
Contact

A further advantage of the BH structure is that, by using a small active region, the gain profile is considerably narrowed. Thus the emission spectrum can consist of a single line – a considerable advantage in long-haul routes operating at a zero dispersion wavelength. Unfortunately the gain profile is dependent on the junction temperature, and so the wavelength of emission can change during operation so introducing dispersion. One solution is to use a Peltier cooler to stabilise operation.

A truly single-mode source results from distributing the feedback throughout the laser, the so-called *distributed feedback, or DFB,* laser. In these devices, a grating replaces the Fabry-Perot cavity resonator (Fig. 3.19). The effect of this grating is to select just one propagating mode. This happens because each perturbation reflects some of the light and, in order to propagate successfully, the phase of the twice reflected light must match that of the incident light. We can write this condition as

$$2n_2 \text{Å} = m\lambda_o \tag{3.121}$$

where Å is the period of the grating, n_2 is the refractive index of the material above the grating, and m is an integer. (The factor of 2 appears in the left-hand side of (3.121) because the light must be reflected twice in order to be in phase with the incident wave.) If (3.121) is not satisfied, the scattered light from the grating will interfere destructively, and the wave will not propagate.

Equation (3.121) is a special case of *Bragg's law* and if m equals unity, the wave is said to be incident at the first Bragg condition. It is also possible for light to be reflected using the second Bragg condition. In fact we can see that if $m = 2$, the grating period will increase, so making it easier to manufacture. We should note that the grating is not part of the active layer. This is because a grating in the active region will cause surface dislocations, and this will increase the non-radiative recombination rate. Instead, the grating is usually placed in a waveguide layer where it interacts with the evanescent field.

A modification of the DFB laser is the *distributed Bragg reflector, DBR,* laser. In this device, short lengths of grating, which act as frequency selective reflectors, replace the Fabry-Perot resonator. Hence many modes propagate in the active region, but only a single wavelength is reflected back and undergoes amplification.

The threshold current of both these devices is typically 20 mA, and their linewidth is quite narrow with linewidths of the order of kHz. Thus high data-rate/long-haul routes often use these devices. As we have seen, these lasers rely on the grating period to select a particular wavelength. However, changes in temperature will cause the grating to expand or contract, and so the wavelength will change. We

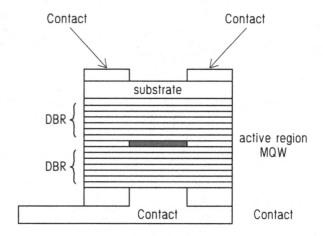

Fig. 3.20 Basic structure of a VCSEL

can control the laser temperature by mounting the semiconductor on a *Peltier cooler*. If we then place a thermistor close to the device, we can use a simple control loop to maintain the laser temperature. Altering the wavelength is very useful particularly when considering heterodyne detection, examined in Chap. 6.

A further laser is the vertical cavity surface-emitting laser or VCSEL. The basic structure of a VCSEL is shown in Fig. 3.20. As can be seen, a VCSEL is similar to a surface-emitting LED in that light is taken from the surface of the device. Of course, there must be optical feedback using mirrors of some sort to be a laser. In the device shown in Fig. 3.20, optical confinement comes from growing DBR mirrors above and below the active region. VCSELs for use in optical fibre links are usually packaged in a TO5 can, sometimes with a small lens to couple into a fibre. The result is a highly collimated beam which is ideal for simple free-space links.

Light emission comes from the active region, and, as we have seen, the wave-length of operation is dependent on the band-gap of the material used. There is an alternative in the form of *quantum wells* in which the wavelength of operation is dependent on the width of the quantum well. For short wavelength operation, the well is fabricated out of GaAs surrounded by AlGaAs walls. For longer wavelengths, InGaAsP wells and InP walls can be used. Quantum wells (Fig. 3.21) are produced by making the active region very thin (<10 nm). At such a thickness, quantum mechanical effects have to be taken into account. Such lasers exhibit very low threshold currents (typically less than 5 mA) and are very efficient. It is possible to increase the power by using multiple quantum wells (*MQW*) and such lasers have been demonstrated with powers in excess of 5 W. MQW are used in DFB, DBR and VCSELs.

3.6   Solid-State and Gas Lasers

So far we have confined our discussion to semiconductor laser diodes. Although we find these devices in optical fibre links, we seldom find them in free-space optical links because of their low output power. Instead, we can use high-power solid-state

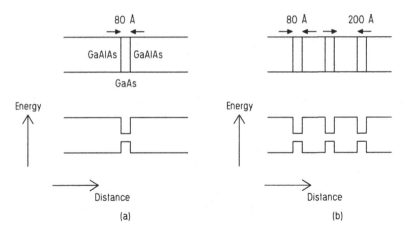

Fig. 3.21 Single and multiple quantum wells

or gas lasers. Such lasers are usually physically large, and modulation of the light output can prove difficult. In spite of this, such lasers are often used in the laboratory, and their application to free-space links is developing.

We begin our study by examining the Nd^{3+}:YAG laser which is extensively used in manufacturing industry as a cutting tool. We will then go on to study the HeNe gas laser, which is often used as a teaching aid in laboratory experiments. We will leave the problem of modulating the light output until we consider light-wave modulation later.

3.6.1 Nd^3 +:YAG Lasers

Neodymium lasers are solid-state lasers in which Nd^{3+} is used as a dopant in the host material. This host material can be certain types of glass, or single crystal rods of yttrium-aluminium-garnet, YAG - $Y_3Al_5O_{12}$, where the Nd^{3+} ions displace some of the yttrium atoms. As Nd^{3+}: YAG is an insulating material, we have to generate a population inversion by external means. The most common way of producing a population inversion is to pump the laser rod with the output of a tungsten-halogen lamp. Under these conditions, we find that several watts of output power are produced for several hundred watts of pump power. Thus we can see that NdH: YAG lasers are very inefficient devices (typically 1 per cent).

Figure 3.22 shows the schematic of a typical Nd^{3+}:YAG laser. The crystal rod is placed inside a Fabry-Perot cavity, to provide optical feedback, and one of the mirrors is made slightly reflecting in order to couple light out of the cavity. We should note that the ends of the crystal rod are cut at an angle to the axis known as the *Brewster* angle. The use of Brewster windows means that only transverse magnetic, *TM*, light is coupled out of the laser rod. This light is reflected off the Fabry-Perot

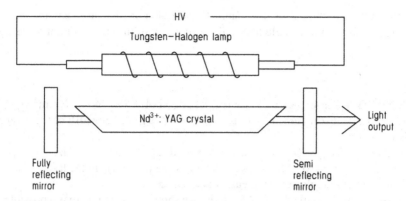

Fig. 3.22 Schematic of a typical Nd^{3+}:YAG laser

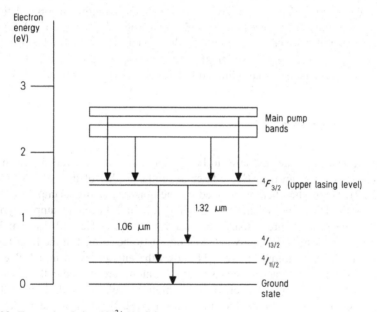

Fig. 3.23 Energy levels in a Nd^{3+}:YAG laser

mirrors, and so the TM light will have a lower threshold than TE light. Linearly polarised light is of great importance if we wish to use the external modulators which we consider in 3.7.3.

As we should expect, the emission wavelength of a Nd^{3+}:YAG laser is dependent on electron transitions between levels in the crystal material, as shown in Fig. 3.23. Electrons appear in the upper lasing level, the $^4F_{3/2}$ level, by dropping down from the main pump bands in the higher F, G and H levels. The spontaneous lifetime of electrons in the upper lasing level is typically 500 μs. In decaying to the lower lasing $^4I_{11/2}$, the electrons lose 1.17 eV, and so the wavelength of the laser is 1.06 μm. By incorporating a highly selective Fabry-Perot etalon in the laser cavity, we can operate the laser with a single longitudinal mode, resulting in operation at a single frequency.

Example

A Nd^{3+}:YAG laser has a cavity length of 20 cm and a refractive index of 1.5. Determine the number of laser modes if the gain linewidth of the laser is 0.674 nm.

By using Eq. (3.101), we find that the mode spacing is 500 MHz. The gain linewidth is 18×10^{10} Hz, and so 360 modes can propagate in the laser cavity. This is why a highly selective etalon is often used.

Before we leave the Nd^{3+}:YAG laser, we should note that there is a possible laser transition from the upper lasing level to the $^4I_{3/12}$ level. Electrons dropping to this level lose 0.941 eV of energy, resulting in emission at a wavelength of 1.32 μm – one of the transmission windows in optical fibre links. We can achieve operation at this wavelength by using coated cavity mirrors that only reflect 1.32 μm light. Although these lasers could be used in optical fibre links, longer wavelengths offer lower attenuation, and so 1.32 μm Nd^{3+}:YAG lasers are usually limited to laboratory experiments.

3.6.2 HeNe Lasers

The HeNe laser was the first laser to be operated continuously rather than pulsed. The schematic diagram of a HeNe laser is similar to that of the Nd^{3+}:YAG laser (Fig. 3.22), except that there is no need for the tungsten-halogen lamp. Instead, we can excite the HeNe by ionising the gas using a high voltage d.c. supply, typically 1–2 kV. The gas mixture usually contains 1 mm Hg of He and 0.1 mm Hg of Ne. Excitation by the HV supply causes a plasma to be formed in the laser tube, so exciting the helium atoms. Figure 3.24 shows the energy levels in a HeNe laser. When a plasma is struck across the laser tube, He atoms are excited to the 2^1s and 2^3s levels. As the He atoms are excited to this higher level, they collide with the Ne atoms, so exciting them to the 2 s and 3 s levels. The Ne electrons then ultimately decay to the ls level, from which they relax to the ground state by collision with the tube walls. The transition to the ls level can occur through several different routes, each one generating light of a specific wavelength. So, although the He atoms are excited by the electrical discharge, it is the Ne atoms that cause the laser output.

The most commonly used transition in HeNe lasers is from the 3s level to the $2p$ level. In doing this, the electrons lose an energy of 1.963 eV which results in light of wavelength 0.633 μm – red light. It is also possible for decay to occur from the 2 s level to the $2p$ level. In this case, the energy difference is 0.816 eV, and so we can see that light is emitted at 1.523 μm. This coincides with the lowest attenuation window in optical fibre, and so we should expect to find 1.523 μm lasers in optical fibre communications links. However, the strength of the 1.523 μm line is very low,

Fig. 3.24 Energy levels in a
HeNe laser

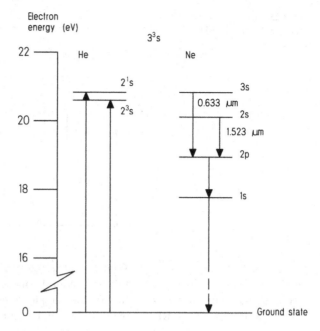

typically <200 μW, and so HeNe lasers are not very useful in optical links. Instead, they are often used as a cheap source of 1.523 μm light for use in the laboratory.

3.7   Light-Wave Modulation

In this section we will consider various techniques for modulating the output of a light source. With a semiconductor light source, such as the SLD or LED, we can modulate the light output by varying the drive current. However, solid-state and gas lasers are usually CW devices, and so we must use some form of external modulator.

We begin by studying LED and SLD drive circuits. We will then go on to examine external modulators, which are used in high-data-rate systems.

3.7.1   LED Drive Circuits

For analogue modulation of an LED, we can use the simple class A amplifier, shown in Fig. 3.25. Provided the modulation depth, m, is less than 100 per cent, no signal distortion will occur. The modulation depth is defined as

$$m = \frac{\delta I}{I_B} \tag{3.122}$$

Fig. 3.25 A simple
analogue driver for LED
sources

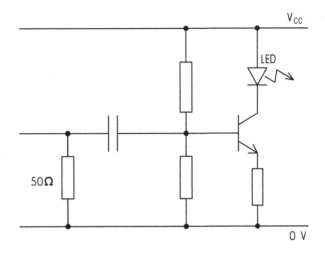

where I_B is the LED bias current. With careful selection of the drive transistor, the time constant of the LED will limit the maximum frequency of operation.

For digital modulation, we can use the simple transistor switch shown in Fig. 3.26a. In this circuit, RL limits the LED current while R_1 limits the transistor current. The purpose of the capacitor, C_1, is to provide a speed-up transient to charge and discharge the LED capacitance. This circuit is suitable for data rates less than 100 Mbit/s.

For operation at data rates greater than 100 Mbit/s, an emitter-coupled driver will often suffice (Fig. 3.26b). When the input is high, T_1 turns on so diverting current away from the LED, which then turns off. When the input is low, T_1 turns off, and the LED turns on.

For commercial applications, most manufacturers supply a package containing the LED and all the drive circuitry. The light output is taken from a short length of fibre, a *fibre pig-tail*, or through a connector housing. Hence, the only connections that need to be made to the unit are the power supply and the signal.

3.7.2 SLD Drive Circuits

The requirement to bias the laser at, or above, threshold means that SLD drive circuits can be complex. In addition, because I_{th} increases with temperature, a feedback loop regulates the diode current. So, a typical SLD drive circuit consists of a constant current source, incorporated in a feedback loop. Such a circuit is shown in schematic form in Fig. 3.27, in which a monitor photodiode attached to the non-emitting laser facet provides the feedback signal. In order to alleviate heatsinking problems, most commercial laser packages incorporate semiconductor Peltier coolers, which also help to keep the threshold current low.

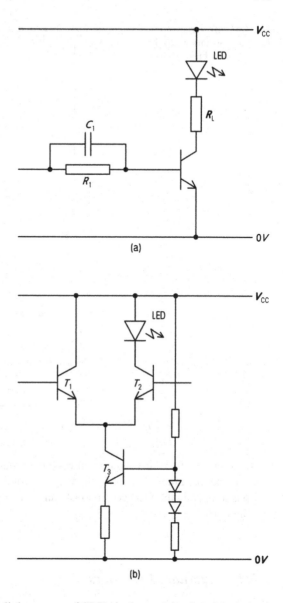

Fig. 3.26 (a) A simple digital driver with speed-up capacitor and (b) a basic emitter-coupled switch for LED light sources

As we have already seen, the light output of SLDs is due to stimulated emission. As this process is faster than spontaneous emission, the emitter-coupled circuit of Fig. 3.26b, shown previously, can be used. However, for high-speed operation, we must specify microwave bipolar transistors, or GaAs MESFETs. Although the rise time of the laser optical pulse can be very fast ($\approx$ 500 ps), the fall time is usually longer, >1 ns. Charge storage in the active region causes this effect, and so the fall time often limits the maximum speed of modulation.

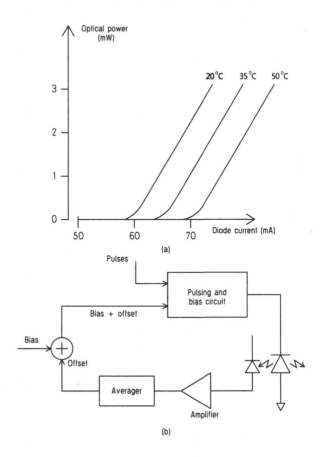

Fig. 3.27 (a) Variation of I_{th} with temperature and (b) a simple laser bias stabiliser

In an optical fibre link, additional dispersion due to laser chirp can also place a limit on the maximum data rate. One way of avoiding both charge storage and laser chirp is to operate the SLD continuously and use an external modulator to modulate the light output.

3.7.3 External Modulators

External modulators fall into two broad areas: waveguide devices for use in optical fibre links and bulk modulators for use in high power free-space links. Bulk modulators come in a variety of forms, and we will consider some of these at the end of this section. Nearly all waveguide modulators are made using lithium niobate, and it is these that we will consider first.

When we examined optical couplers in the previous chapter, we briefly considered a single-mode, waveguide coupler fabricated on a lithium niobate, $LiNbO_3$, substrate. One property of this material is that the refractive index varies according to the strength of an externally applied electric field, the so-called *electro-optic* effect. We can exploit this effect to produce phase and intensity modulators. Although we are more usually concerned with varying the intensity of a light source rather than the phase, we will initially consider phase modulators. This will help us when we come to examine the intensity modulator.

Figure 3.28a shows the basic structure of a typical phase modulator. In this device, the electrodes either side of the waveguide set up an electric field, E, across the guide. This has the effect of increasing the refractive index and hence the propagation time and phase. We can write the change in phase experienced by the optical signal $\delta\phi$ as

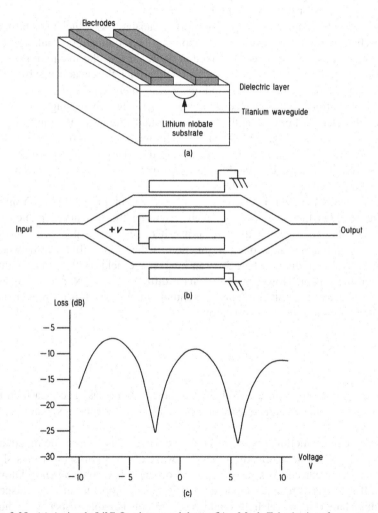

Figure 3.28 (**a**) A simple LiNbO₃ phase modulator; (**b**) a Mach-Zehnder interferometer; and (**c**) measured transfer function of a Mach-Zehnder interferometer

$$\delta\varnothing = \frac{2\pi}{\lambda_o}\delta nL \tag{3.123}$$

where L is the length of the guide. The change in refractive index, δn, is related to the electric field by

$$\delta n = n^3 \frac{r}{2}E \tag{3.124}$$

and so the phase change is directly proportional to the applied voltage. The parameter r is called the *electro-optic coefficient*. In LiNbO₃, $r = 30 \times 10^{-12}$ m/V and $n = 2.2$, and so a typical field of 10^7 V/m results in $\delta n = 1.6 \times 10^{-3}$. If the substrate is GaAs, then δn, for the same field, is 2.57×10^{-4}.

The electrodes in the phase modulator form a capacitor, C, which must be charged and discharged by the voltage source supplying the modulating signal. The output resistance of this source is usually fixed at 50 Ω, and so the maximum operating speed is set by the time constant 50C. We could reduce the capacitance by decreasing the length of the electrodes and increasing their separation. Unfortunately increasing electrode length reduces the phase shift, while increasing the separation reduces the electric field strength. One way round this is to use travelling-wave electrodes as shown in Fig. 3.28a. In order to eliminate electrical reflections, we must make sure that the whole system is matched to the characteristic impedance of the transmission line. With this modification, the maximum operating speed can be greater than 50 GHz.

In the Mach-Zehnder interferometer [5] shown in Fig. 3.28b, a Y-junction waveguide splits the input power equally between the two arms of the device. Phase modulators placed in the two arms alter the relative phases of the fields prior to recombination in another Y junction. If the phase difference between the two paths is $2N\pi$ radians, where N is an integer, the fields will add and light will appear at the output. However, if the phase difference is $(2N + 1)\pi$ radians, the waves will cancel each other out and the output will be zero. It is a simple matter to show that the output power is given by

$$P_{\text{out}} = P_{\text{in}} \cos^2\left(\frac{\Delta\phi}{2}\right) \tag{3.125}$$

where $\Delta\phi$ is the phase difference between the two branches. The measured transmission/drive voltage characteristic of a typical Mach-Zehnder modulator is shown in Fig. 3.28c.

Mach-Zehnder modulators are of great use when a laser has to be modulated at high speed. As we have already noted, the wavelength of emission varies slightly when the drive current to a laser is pulsed on and off (a phenomenon known as chirp). If the laser is a single-mode device designed to operate in a low dispersion link, this change in wavelength could result in considerable dispersion. Thus for high-speed operation, we can operate the laser continuously and use a Mach-

Zehnder modulator to modulate the output. Such a technique is widely used at present.

If we want to modulate the light output of a high-power laser, we have to use bulk modulators. This is because the optical power density in a single-mode waveguide modulator would be extremely high and could cause damage to the device. However, if we have a wide output beam, the power density will be lower, and so modulator damage is less likely to occur. In the main, we find that bulk modulators fall into two groups: those that use the electro-optic effect and those that use the acousto-optic effect.

Bulk modulators using the electro-optic effect rely on the polarisation changes that the modulator crystal produces when excited. Figure 3.29a shows the schematic of a typical electro-optic amplitude modulator. Let us initially neglect the effect of the quarter-wave retardation plate. As we can see, a polariser (which is aligned to the crystal lattice) is used at the input to the device. As the light passes through the energised crystal, two things occur. Firstly the polarisation is altered by a maximum

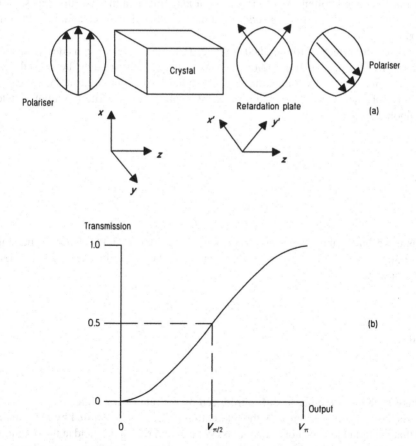

Fig. 3.29 (a) Schematic diagram of an electro-optic bulk modulator and (b) transfer function of the modulator

of $\pi/2$, so that the light travels along the x'- and y'-axes. The second effect is that the light travelling along the x'-axis experiences a maximum phase delay of $\pi/4$ radians with respect to the y'-axis. This effect is known as *birefringence,* which we can interpret as the x' wave travelling in a higher refractive index material than the y' wave. (We should note that both the polarisation shift and the phase shift depend on the voltage across the crystal.)

At the output of the crystal, the light passes through an output polariser which is aligned at right angles to the input polariser. (We are temporarily neglecting the effects of the retardation plate.) Thus we can see that if the crystal is energised, the total change in polarisation is $\pi/2$ radians, and light passes through the modulator. However, if the applied voltage is zero, the polarisation state is maintained, and no light appears at the modulator output. If we want to modulate the light with a sinusoidal waveform, we need to bias the crystal so that it introduces a $\pi/4$ polarisation shift. (This enables us to vary the polarisation shift between 0 and the maximum $\pi/2$ radians.) Rather than supply a fixed bias to the crystal, a quarter-wave retardation plate is often used to introduce a $\pi/4$ shift regardless of the bias signal. Thus we can see that the energised crystal only needs to introduce a maximum shift of $\pi/4$ radians.

To see the effect of the applied voltage on the light, let us consider vertically polarised light, propagating along the z-axis. In passing through the crystal, the polarisation changes by a maximum of $\pi/4$ radians. Thus the light in the crystal is equally split between the x'- and y'-axes. We can therefore write the input E field components as

$$E_{x'}(0) = A$$

and

$$E_{y'}(0) = A$$

where we have taken the phasor representation of the E fields. Now, in passing through the modulator, the x' wave experiences a total phase shift of Γ radians. Thus we can write

$$E_{x'}(l) = A \exp(-j\Gamma)$$

and

$$E_{y'}(l) = A$$

where Γ includes the phase shift introduced by the retardation plate.

As these waves pass through the retardation plate, they are shifted by $\pi/4$ radians so that the light appears along the y-axis. We can find the total E field at the output of the modulator by taking the vector sum of these components. Thus

$$E_{yo} = -\frac{A}{\sqrt{2}} \exp(-j\Gamma) + \frac{A}{\sqrt{2}}$$
$$= \frac{A}{\sqrt{2}}[1 - \exp(-j\Gamma)]$$

Now, the intensity of the output light is given by

$$I_o = E_{yo}E_{yo}{}^*$$
$$= \frac{A^2}{2}|[1 - \exp(-j\Gamma)][1 - \exp(j\Gamma)]|$$
$$= A^2(1 - \cos\Gamma)$$
$$= 2A^2 \sin^2(\Gamma/2)$$

and the intensity of the input light is

$$I_i = |E_{x'}(0)|^2 + |E_{y'}(0)|^2$$
$$= 2A^2$$

Thus we can write the modulator transfer function as

$$\frac{I_o}{I_i} = \sin^2(\Gamma/2)$$

With Γ given by

$$\Gamma = \frac{2\pi l}{\lambda_o}\delta n$$

where δn is defined by (3.124). As we are considering a longitudinal field, we can write

$$\Gamma = \frac{\pi}{\lambda_o}n^3 rV$$

If we also define V_p as

$$V_\pi = \frac{\lambda_o}{n^3 r}$$

where V_π is the voltage needed to produce a phase shift of 7 t radians, we get

$$\frac{I_o}{I_i} = \sin^2\left|\frac{\pi}{2}\frac{V}{V_\pi}\right| \tag{3.126}$$

From (3.126) we can see that the transfer function follows a $\sin^2$ form as shown in Fig. 3.29b. From this figure we can see that the region around the 50 per cent transmission is reasonably linear, and this is where the modulator is normally operated. If we use a quarter-wave retardation plate after the crystal, the modulator will operate in the middle of the linear region resulting in minimal distortion of the modulating signal. A modification to the modulator we are considering is to place the electrodes transverse to the crystal. Under these circumstances, the voltage required to produce a $\pi/4$ phase shift is considerably reduced. At present, the maximum speed of these devices is typically 2 GHz.

Example
A bulk modulator uses potassium dihydrogen phosphate, KH_2PO_4 also known as *KDP*. The refractive index of this material at 0.633 μm is approximately 1.5, and the electro-optic coefficient, also at the same wavelength, is 11×10^{-12} m/V. Determine the voltage required to produce a phase shift of π radians.
As V_π is given by

$$V_\pi = \frac{\lambda_o}{n^3 r}$$

we find that V_π is 17 kV. With retardation plates this means that we must apply 8.5 kV to give a phase shift of π radians. Such high voltages are typical for the type of modulator we are considering. In view of the magnitude of the drive voltage, we can see that the design of the modulator driver is by no means easy!

We can reduce this voltage to a more manageable one by using *transverse electrodes*. If we use a crystal that is 10 cm long, with a width of 1 cm, we get a V_π voltage of 1.7 kV. As we have already seen, lithium niobate also exhibits electro-optic properties. In $LiNbO_3$, the refractive index at 0.633 μm is 2.2, and r is 31×10^{-12} m/V. Thus we find that V_π is 192 volt, for a transverse modulator with a crystal of the same dimensions.

The other type of bulk modulator uses the photo-elastic effect. Figure 3.30a shows the schematic of a typical acousto-optic modulator, *AOM*. The transducer at the bottom of the modulator launches a travelling acoustic wave into the bulk material. As the crests of the sound wave propagate through the crystal, they cause the refractive index of the material to increase due to stress. Thus a travelling wave of high and low refractive index is set up in the crystal. These areas of high refractive index act as mirrors, and so any light incident on the crystal is deflected as shown in Fig. 3.30b.

(continued)

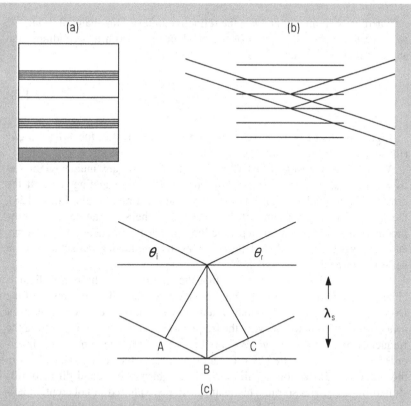

Fig. 3.30 (a) Schematic of an acousto-optic modulator; (b) deflection of beam from the crystal; and (c) reflection of beam by area of high refractive index

Figure 3.30c shows the deflection of a light ray off a high refractive index region. In order for the light to propagate successfully after deflection, the path length AB + BC must be an integral number of optical wavelengths in the crystal. So we can write

$$2\lambda_s \sin\theta_r = \frac{m\lambda_o}{n}$$

or

$$\theta_r = \sin^{-1}\left[\frac{m\lambda_o}{n}\frac{1}{2\lambda_s}\right] \qquad (3.127)$$

(continued)

where m is an integer and n is the refractive index of the material. If $m = 1$, we have *first-order diffraction* of the incident light. With this condition, the angle of incidence equals the angle of reflection, and so

$$\theta_r = \sin^{-1}\left[\frac{\lambda_o}{n}\frac{1}{2\lambda_s}\right] \tag{3.128}$$

Light rays that satisfy (3.128) are said to be incident at the *Bragg angle*. (Indeed, an alternative name for an AOM is a *Bragg cell*.)

We can use the result of (3.127) to deflect the path of any incident light. Let us assume that a carrier wave of frequency f_c is propagating through the crystal. If the incident light hits the crystal at the Bragg angle, Eq. (3.128), the first-order output beam will be reflected at the same angle. If we now modulate the carrier with a step function, the carrier frequency will change, and the output beam will be deflected. Thus the modulator can act as a beam deflector, or as a simple switch.

One useful feature of AOMs is that the frequency of the output light is either increased or decreased by the acoustic frequency. This is a result of the Doppler shift, which we encounter whenever a sound source moves towards or away from us. Thus we can vary the frequency of a light source by varying the frequency of the carrier wave. We should also note that the light in the higher-order diffractions is shifted by different multiples of the carrier. Thus, light in the first-order diffraction is shifted by $\pm f_c$, light in the second diffraction is shifted by $\pm 2f_c$, and so on. This feature can be exploited in coherent experiments in the laboratory. Coherent detection uses a local oscillator laser that operates at a different frequency to the source. When the source and local oscillator signals are mixed together, they produce an i.f. equal to the difference between the optical frequencies. In the laboratory, we can generate the local oscillator frequency by passing some of the source light through a Bragg cell. If the frequency of the acoustic signal is that of the i.f., we have no need to use a separate local oscillator laser. Unfortunately, the available power in the higher-order diffractions is quite low, and so the first-order diffraction is most commonly used.

The maximum operating speed of Bragg cells is reached when the acoustic wavelength in the crystal equals the diameter of the light beam, d. Thus we can write

$$f_{max} = \frac{v_{ac}}{d} \tag{3.129}$$

where v_{ac} is the velocity of the sound wave in the crystal. Thus we can see that it is important to reduce the spot size of the incident light. Unfortunately this will increase the power density of the light wave, and so crystal damage may occur.

Example
Light of wavelength 0.633 μm, and 200 μm spot size, is incident on a LiNbO 3 crystal. The acoustic velocity of the material is 6.6 × 10³ m/s. Determine the maximum modulating frequency.

By using (3.129), we get a maximum modulating frequency of 33 MHz. If we reduce the spot size to 50 μm, we get $f_{max} = 130$ MHz which is typical of such devices.

In view of the low bandwidth of these devices, they are seldom used in optical fibre links.

3.8 Fibre Lasers

Considerable work has been carried out in the area of *rare-earth doped silica fibre*. If we dope silica fibre with ions of the rare earth erbium, Er^{3+}, we find that the Er^{3+} ions absorb light at 980 nm and 1.49 μm wavelengths and generates 1.55 μm wavelength light. (This is similar in operation to the Nd^{3+}:YAG laser we encountered in Sect. 3.6.1.) This is a very important result because light is generated in fibre, and so there are no coupling components to be used.

Figure 3.31 shows the schematic of a rare-earth doped fibre laser. Also shown in Fig. 3.31 is the mechanism by which light is generated. Operation is very similar to that of a semiconductor laser; a pump (an 820 nm laser) excites the Er^{3+} electrons to a high energy state from which they rapidly drop to the required level. A population inversion is generated, and optical feedback is provided by using a grating at each

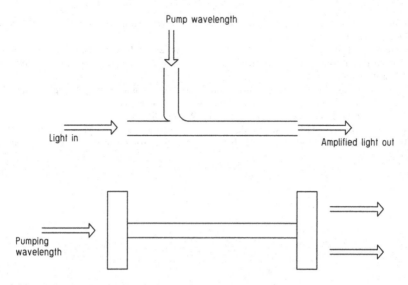

Fig. 3.31 (a) Schematic of a fibre laser and (b) energy levels in the laser

end. This is effectively a DBR laser with the doped fibre taking the place of the semiconductor. Stimulated emission occurs in the region of 1.55 µm.

Problems

1. The wavelength of emission of light from GaAs can be adjusted by adding Al. Determine the proportion of Al to Ga required to emit light at 850 nm.

   [0.035]

2. Sources operating at 1300 and 1550 nm use an InGaAsP alloy with the proportion of Ga to In being adjusted to vary the wavelength. Determine the proportion of Ga for 1300 and 1500 nm.

   [0.253 for 1300 nm and 0.392 for 1550 nm]

3. In a sample of InGaAs, the effective electron mass is $3.7 \times 10{-32}$ kg and that of holes is $1.8 \times 10{-31}$ kg. Estimate the density of carriers in the conduction and valence bands and the total carrier density. Take Eg to be 0.75 eV.

   [2.05×1023 m-3, 2.20×1024 m-3 and 3.41×1017 m-3]

4. The InGaAs sample of question 3 is used in a p + n diode in which the p-type is doped to 1027 and the n-type to 1025 m-3. Determine the barrier potential.

   [0.975 V]

5. The active region of a 1550 nm stripe contact laser has facets with a reflectivity of 0.6 and 0.9, a loss per cm of 10, a length of 300 µm, an active region width of 5 µm, a spontaneous emission linewidth of 10 nm and $\varepsilon r = 5.6$. Determine the gain required for lasing and the threshold current.

   [20 cm-1, 8.4 mA]

Recommended Reading

1. Numai, T., (2015), 'Fundamentals of semiconductor lasers.' Springer Series in Optical Science, Springer Nature, Switzerland
2. Baranov A, Tournie E (eds) (2013) Semiconductor lasers: fundamentals and Applications, Woodhead publishing series in electronic and optical materials. Woodhead Publishing, Cambridge, UK
3. Casey HC, Panish MB (1978) Heterostructure lasers, part A: fundamental principles and part B: materials and operating characteristics. Academic Press, New York
4. Yariv A (1991) Optical electronics, 4th edn. HRW Ltd, Orlando
5. Nayar BK, Booth RC (1986) An introduction to integrated optics'. Br Telecom Technol J 3:5–15
6. Brierley MC, France PW (1987) Neodynium doped fluorozirconate fibre laser. Electron Lett 23:815–817

Chapter 4
Photodiodes

In order to convert the modulated light back into an electrical signal, we must use some form of photodetector. As the light at the end of any optical link is usually of very low intensity, the detector has to meet a high-performance specification: the conversion efficiency must be high at the operating wavelength; the speed of response must be high enough to ensure that signal distortion does not occur; the detection process should introduce the minimum amount of additional noise; and it must be possible to operate continuously over a wide range of temperatures for many years. A further obvious requirement for optical fibre links is that the detector size must be compatible with the fibre dimensions.

4.1   V-I Characteristics of Photodiodes

At present, reverse-biased p-n photodiodes are used as photon detectors in optical communications. In these devices, the semiconductor material absorbs a photon of light, which excites an electron from the valence band to the conduction band. (This is the exact opposite of photon emission which we examined in the previous chapter.) The photo-generated electron leaves behind it a hole, and so each photon generates two charge carriers. This increases the material conductivity, so-called photoconductivity, resulting in an increase in the diode current.

We can modify the familiar diode equation to give

$$I_{\text{diode}} = (I_{\text{d}} + I_{\text{s}})(\exp\left[\text{Vq}/\eta kT\right] - 1) \tag{4.1}$$

where I_{d} is the *dark current*, that is, the current that flows when no signal is present, and I_{s} is the photo-generated signal current due to the incident optical signal. Figure 4.1 shows a plot of this equation for varying amounts of incident optical power. As we can see, there are three distinct operating regions: forward bias, reverse bias and avalanche breakdown. Under forward bias, region 1, a change in

© Springer Nature Switzerland AG 2020   153
M. Sibley, *Optical Communications*, https://doi.org/10.1007/978-3-030-34359-0_4

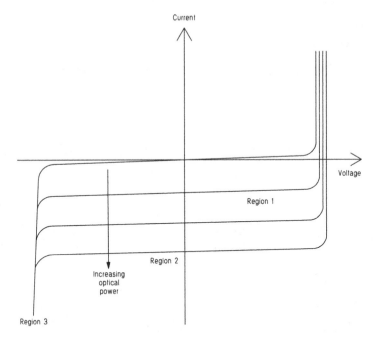

Fig. 4.1 *V-I* characteristic of a photodiode, with varying amounts of incident optical power

incident power causes a change in terminal voltage, the so-called photovoltaic mode. If we operate the diode in this mode, the frequency response of the diode is poor and so photovoltaic operation is rarely used in optical links.

If we reverse bias the diode, region 2, a change in optical power produces a proportional change in diode current. This is the *photoconductive mode* of operation which most detectors use. Under these conditions, the exponential term in (4.1) becomes insignificant, and the reverse bias current is given by

$$I_{\text{diode}} = I_{\text{d}} + I_{\text{s}} \tag{4.2}$$

We can define the *responsivity* of a photodiode, Ro, as the change in reverse bias current per unit change in optical power and so efficient detectors need large responsivities.

Avalanche photodiodes, APDs, operate in region 3 of the *V-I* characteristic. When biased in this region, a photo-generated electron-hole pair, *EHP*, causes avalanche breakdown, resulting in a large diode current for a single incident photon. Because APDs exhibit carrier multiplication, they are usually very sensitive detectors. Unfortunately the *V-I* characteristic is very steep in this region, and so the bias voltage must be tightly controlled to prevent spontaneous breakdown.

Before we go on to examine the structure and properties of PIN and APD detectors, it will be useful to discuss photoconduction in semiconductor diodes. Although most of our discussion will centre around silicon, we can apply the same basic arguments to other materials.

4.2 Photoconduction in Semiconductors

4.2.1 Photon Absorption in Intrinsic Material

As we saw in the last section, when a semiconductor absorbs a photon, an electron is excited from the VB to the CB so causing an increase in conductivity. If the VB electron energy is E_1 and the CB energy level is E_2, then we can relate the change in energy, $E_2 - E_1$, to the wavelength of the incident photon by

$$\lambda_o = \frac{hc}{E_2 - E_1} \tag{4.3}$$

Now, the lowest possible energy change is the band-gap of the material, and so this results in a cut-off wavelength beyond which the material becomes transparent. These cut-off wavelengths are identical to the emission wavelengths of sources made of the same material (see Table 3.1). Hence, silicon responds to light of wavelengths up to 1.1 µm, whereas germanium photodiodes operate up to 1.85 µm. (It may be recalled from our discussion of photon emission in Chap. 3 that sources are made out of direct band-gap materials. However, detectors can be made out of indirect band-gap materials such as Si or Ge.)

The *absorption coefficient*, α, is a measure of how good the material is at absorbing light of a certain wavelength. As light travels through a semiconductor lattice, the material absorbs individual photons, so causing the intensity of the light (the number of photons per second) to fall. The reduction is proportional to the distance travelled and so, if the intensity reduces from I to $I - \delta I$ in distance δx, we can write

$$\frac{\delta I}{I} = -\alpha \delta x \tag{4.4}$$

We can find the intensity at any point in the lattice by taking the limit of (4.4) and integrating. So, if I_0 is the intensity at the surface, $x = 0$, we can write

$$\ln \frac{I}{I_o} = -\alpha x$$

which gives

$$I = I_o \exp(-\alpha x) \tag{4.5}$$

(We should note that the optical power follows an identical exponential decay.) From our discussion of photo-emission, we can intuitively reason that α will vary with wavelength. Figure 4.2 shows this variation for several semiconductor materials. These plots clearly show an absorption edge which is in close agreement with the cut-off wavelength found from (4.3). The variation of gradient with wavelength

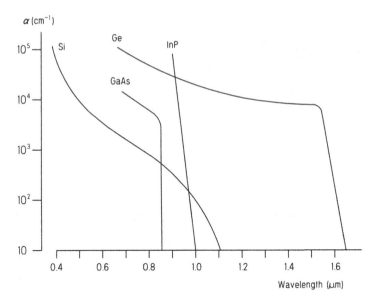

Fig. 4.2 Variation of absorption coefficient with wavelength, for a number of semiconductor materials

is due to the differing density of energy levels in the VB and CB of the material – the same mechanism that causes the spread of wavelength in a semiconductor source.

The absorption coefficient is a very important parameter when considering the design of photodiodes. If the absorbing layer is too thin, a large proportion of the incident light passes straight through, resulting in a low conversion efficiency. If the layer is too thick, the transit time of the carriers is large, limiting the speed of response. Thus there is a trade-off between conversion efficiency and speed of response.

Example
A 40-μm-thick silicon detector has an α value of 6.3×10^4 m^{-1} when receiving 850 nm wavelength light. Determine the proportion of light that is not absorbed, assuming that no reflection takes place from the surface of the material.
We can use (4.5) to give
$$\frac{I}{I_o} = \exp(-\alpha x)$$
$$= \exp(-6.3 \times 10^4 \times 40 \times 10^{-6}) = 0.08$$

(continued)

Thus we can see that only 8 per cent of the light is not absorbed by the detector. This does, however, assume that no light is reflected from the surface of the material. In practice *anti-reflection* coatings are used to minimise reflection from the surface. (Such coatings are similar to the anti-reflection coatings used on spectacles.)

As we shall see later, the carrier transit time of this device is of the order of 100 ps, and so the detector is both efficient and fast.

4.2.2 Photon Absorption in Reverse-Biased p-n Diodes

When a photon of light is absorbed by a semiconductor crystal, an electron is excited to the CB where it takes part in the conduction of current. Thus the reverse bias current consists of the normal leakage current that flows even when there is no incident light, the so-called dark current (I_d), and the photon generated light, the *signal current* (I_s). Hence we can write the total diode current as

$$I_{\text{diode}} = I_d + I_s \tag{4.6}$$

Let us initially examine the dark current term. In the previous chapter, we found that a junction diode under zero bias has a depletion region either side of the junction. Under reverse bias conditions, the external bias voltage has the same polarity as the built-in barrier potential. Thus the depletion region expands and this reduces the current flow. As we saw in the previous chapter, the zero-bias barrier potential is given by (see Eq. (3.19) given previously)

$$V_b = \frac{kT}{q} \ln \left(\frac{N_d N_a}{n_i^2} \right) \tag{4.7}$$

We also found that the barrier potential is given by (see Eq. (3.27) given previously)

$$V_b + V_r = \frac{q}{2\varepsilon} \left(N_d w_{nd}^2 + N_a w_{pd}^2 \right) \tag{4.8}$$

where we have modified (3.27) by the addition of the external bias voltage V_r. Thus we can see that the depletion region width depends on the reverse bias.

By following a similar analysis to that used in Sect. 3.1.4, we find that the dark current is given by

$$I_d = I_0 (\exp (qV/kT) - 1) \tag{4.9}$$

where V is the total reverse bias voltage and

$$I_o = \left(\frac{D_n q n_p}{x_p} + \frac{D_p q p_n}{x_n} \right) \times \text{Area}$$

Let us now turn our attention to the illuminated diode. When a semiconductor diode absorbs a photon, an EHP is produced which increases the density of charge carriers. Thus the leakage current, as given by (4.9), is effectively increased. Now, the signal current, I_s, is directly dependent on the rate of generation of EHPs which, in turn, is dependent on the number of incident photons per second. With an incident optical power P consisting of photons of energy E_{ph}, the number of photons per second is

$$\begin{aligned} N_{ph} &= \frac{P}{E_{ph}} \\ &= \frac{P \lambda_o}{hc} \end{aligned} \tag{4.10}$$

Only some of these photons generate electron-hole pairs. Specifically, the number of carrier pairs generated per second, N, is given by

$$\begin{aligned} N &= \eta N_{ph} \\ &= \frac{\eta P \lambda_0}{hc} \end{aligned} \tag{4.11}$$

where η is known as the *quantum efficiency*. From our previous discussion, it should be clear that η is highly dependent on α. However, as we shall see presently, η is also dependent on the device structure.

Now, the photo-generated current is equal to the rate of creation of extra charge. Thus I_s will be given by

$$\begin{aligned} I_s &= qN \\ &= q \frac{\eta P \lambda_0}{hc} \end{aligned} \tag{4.12}$$

We can rearrange this equation to give the change in current per unit change in optical power, the *responsivity*. Hence

$$R_o = \frac{I_s}{P} = \frac{q \eta \lambda_0}{hc} \tag{4.13}$$

As we can see, R_o is directly proportional to λ_o, and Fig. 4.3 shows the theoretical variation of R_0 with λ_o for various values of η. Also shown is the responsivity characteristic of a typical Si photodiode. At long wavelengths, the curve shows a

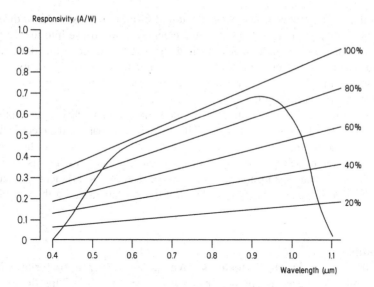

Fig. 4.3 Variation of responsivity with wavelength for a typical Si PIN photodiode. Also shown is the theoretical variation of R_o for a range of quantum efficiencies

Fig. 4.4 Schematic of a reverse-biased p-n junction photodiode

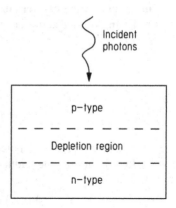

sharp cut-off coinciding with the cut-off wavelength for silicon. However there is also a lower cut-off region. To explain this we have to examine the structure of a reverse-biased p-n photodiode. Most photodiodes have a planar structure, and so any incident light must first pass through the p-type region before reaching the depletion layer. Because of this, absorption can occur in either the p-type region, the depletion region or the n-type region (Fig. 4.4). As the light intensity in the n-type is very low, we can generally ignore absorption in this region. However, absorption in the p-type has a dramatic effect on the quantum efficiency.

Incident light of a low wavelength gives a low penetration depth ($1/\alpha$), resulting in EHP generation in the p-type layer. Unless carrier generation occurs within a diffusion length of the depletion layer boundary, the EHPs recombine, and the diode current does not change. However, if photon absorption occurs within a diffusion length of the depletion layer boundary, the electrons will diffuse into the depletion region. As this is an area of high electric field, they are swept across the diode, and so the diode current increases. Thus useful absorption in the p-type only occurs within a diffusion length of the depletion layer. This explains why the quantum efficiency reduces with low wavelength.

Let us now consider absorption in the depletion region. Photon absorption in this region causes the photo-generated EHPs to be swept apart by the electric field – electrons to the n-type and holes to the p-type. The carriers increase the majority carrier density in these regions, and so the diode current increases. This is obviously more efficient than absorption in the p-type. Hence an efficient photodiode should have a thin p-type layer, less than a diffusion length and a thick depletion region.

Example
A silicon p^+n junction diode is formed from p-type material with $N_a = 10^{24}$ m^{-3} and n-type material with $N_d = 10^{21}$ m^{-3}. The diode has a reverse bias voltage of 20 volts across it. Determine the width of the depletion region and the maximum field strength. (The density of thermally generated carriers in silicon is 1.4×10^{16} m^{-3} and $\varepsilon_r = 11.8$.)

From (4.7) the zero-bias barrier potential is

$$V_b = \frac{kT}{q} \ln \left(\frac{N_d N_a}{n_i^2} \right)$$
$$= 25 \times 10^{-3} \ln \left(5.1 \times 10^{12} \right)$$
$$= 0.73 \text{ Volt}$$

Thus the total barrier potential is 20.73 volt.
In order to find the width of the depletion region, we must apply (4.8).

$$V_b + V_r = \frac{q}{2\varepsilon} \left(N_d w_{nd}^2 + N_a w_{pd}^2 \right)$$

and so

$$20.73 = 7.7 \times 10^{11} w_{nd}^2 + 7.7 \times 10^{14} w_{pd}^2$$

Now, as the total charge in the depletion region equals zero, we have

$$w_{pd} N_a = w_{nd} N_d$$

Therefore

(continued)

$$20.73 = 7.7 \times 10^{11} w_{nd}{}^2 + 7.7 \times 10^8 w_{nd}{}^2$$

giving $W_{nd} = 5.2\ \mu m$ and $W_{pd} = 5.2$ nm

Thus we can see that the width of the depletion layer is approximately 5.2 μm, and it is mainly in the lightly doped n-type material.

The maximum field strength occurs at the junction between the two materials. Thus

$$
\begin{aligned}
E_{max} &= -\frac{dV}{dx_{x=0}} \\
&= \frac{qN_d}{\varepsilon} w_{nd} \\
&= 8\ \text{MV/m} \\
&= 8\ \text{V/\mu m}
\end{aligned}
$$

The diode is now illuminated with 100 nW of 850 nm light. If the p-type is 10 μm thick, and the n-type is 400 μm thick, determine the light intensity in the depletion region. (Assume that the diode is anti-reflection coated and that the reflectivity of this coating is 20 per cent at 850 nm.) Take $\alpha = 6.3 \times 10^4\ \text{m}^{-1}$.

As the reflectivity of the diode is 20 per cent, only 80 per cent of the incident light will penetrate to the p-type. Thus the optical power at the surface of the p-type is

$$I_p = 80\ \text{nW}$$

This light will be exponentially attenuated as it crosses the diode. The p-type is 10 μm thick, and the depth of the depletion region in the p-type is negligible in comparison. Thus the intensity at the depletion region boundary is

$$
\begin{aligned}
I_{dep} &= 80 \times 10^{-9} \exp\left(-6.3 \times 10^4 \times 10 \times 10^{-6}\right) \\
&= 43\ \text{nW}
\end{aligned}
$$

As the depletion region is 5.2 μm thick, the power at the n-type edge of the depletion region is 31 nW. Thus only 12 per cent of the optical power contributes to EHP generation, that is, $\eta = 12$ per cent. In practice, the situation is not as bad as this.

As we have already discussed, if carriers are generated within a diffusion length of the depletion region, they will contribute to the diode current.

(continued)

If we take $D_p = 13 \times 10^{-4}$ m^2 s^{-1}, $D_n = 50 \times 10^{-4}$ m^2 s^{-1} and $\tau_n = \tau_p = 10$ μs, we find

$$L_p = 114 \text{ μm and}$$

$$L_n = 223 \text{ μm}$$

As we can see, the depth of the p-type is less than a diffusion length, and so absorption in this region will produce useful EHPs. Thus useful absorption takes place over

$$10 + 5.2 + 223 = 238.2 \text{ μm}$$

The light power after travelling this distance is

$$80 \times 10^{-9} \exp\left(-6.3 \times 10^4 \times 238.2 \times 10^{-6}\right) = 0.024 \text{ pW}$$

and so we can see that almost all the light that penetrates through the anti-reflection coating generates EHPs. Thus the quantum efficiency is approximately 80 per cent.

This example has shown that, although the depletion region can be small, the efficiency of the diode can be quite high due to absorption within a diffusion length of the depletion region. However, a thin depletion region implies a large depletion capacitance which results in a slow detector. In addition, the diffusion of carriers to the depletion region is a slow process, and this reduces the detector speed. The solution is to use a PIN photodiode, and this is the subject of the next section.

4.3   PIN Photodiodes

As we have just seen, a simple p$^+$n diode can act as an efficient detector. However, the width of the depletion region results in a high diode capacitance. The solution is to use a PIN structure in which, under reverse bias conditions, the thickness of the depletion region is effectively that of the intrinsic material.

Figure 4.5 shows the schematic diagram of a PIN photodiode. Also shown is the variation of the reverse bias electric field intensity, E, across the diode. As can be seen, the field reaches a maximum in the intrinsic layer, the *I-layer*, which is usually high enough to enable the carriers to reach their saturation velocity. As a result, PIN photodiodes are usually very fast detectors. In the figure, we have labelled the I-layer as *n*-. We use this notation to show that the material has been lightly doped with about 10^9 donor atoms per cubic metre. This is done because it is difficult to produce

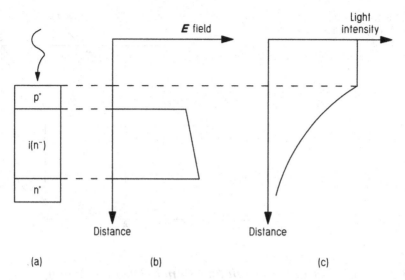

Fig. 4.5 (**a**) PIN photodiode schematic, (**b**) electric field intensity and (**c**) light intensity across the photodiode

a totally intrinsic layer, and so the doping controls the diode characteristics. Because of this doping, the p- and n-type regions are heavily doped with $p^+ \approx 10^{24}$ m^{-3} and $n^+ \approx 10^{22}$ m^{-3} in order to approximate to a PIN diode.

4.3.1   Structure

Figure 4.6 shows a typical Si PIN photodiode structure. The diameter of these devices ranges from 50 μm, for high-speed operation, to 200 μm, for low-speed operation. The greater the diode diameter, the greater the light-collecting capability; however, high-speed operation requires a small detector. In order to avoid this problem, some photodiode packages have a hemi-spherical lens which collects light from a large area and focuses it on to a small area detector.

Under reverse bias conditions, all the n$^-$ carriers are swept away, and so the depletion region extends from the p-type right through to the n-type. The bias voltage at which this occurs is known as the *punch-through* voltage. If the bias increases beyond this point, the depletion region will extend beyond the contact rim. Since the SiO$_2$ layer is transparent to light, the semiconductor can absorb photons that do not pass through the p + layer.

This absorption mechanism is more efficient than absorption through the p-type layer, and so the overall quantum efficiency can be as much as 85 per cent for infrared light.

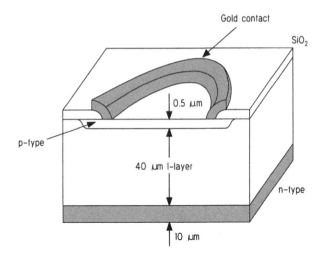

Fig. 4.6 Cross-section through a typical silicon PIN photodiode

4.3.2   Depletion Layer Depth and Punch-Through Voltage

In a normal p-n diode, the depletionregion extends into both the p- and n-type layers. However, as we have already seen, if the doping in the p-type is higher than in the n-type, giving a p^+-n^- diode, then most of the depletion region will be in the n^- material. As PIN diodes have a p + n-n + structure, nearly all of the depletion region exists in the lightly doped intrinsic layer. If the doping level in the I-layer is N_d, then, by applying Eq. (4.8), we can approximate the width of the depletion layer by

$$d = \left(\frac{2\varepsilon_o \varepsilon_r (V_b + V_r)}{q N_d} \right)^{\frac{1}{2}} \tag{4.14}$$

where V_b is the zero-bias barrier potential ($= 0.75$ V for Si). If the I-layer is completely depleted – the *punch-through condition* – the depletion layer depth will be that of the I-layer. So, we can find the punch-through voltage by rearranging (4.14) to yield V_r.

Example
A Si PIN photodiode has a 40-μm-thick I-layer with N_d equal to 10^{19} m^{-3}.
Determine the punch-through voltage.
   By applying (4.14) we find

$$V_b + V_r = \frac{q d^2}{2\varepsilon_o \varepsilon_r} N_d$$
$$= 12.25 \text{ Volt}$$

Thus the punch-through voltage for this diode is 11.5 Volt.

As mentioned previously, the electric field intensity is usually made high enough to ensure that the carriers drift at their saturation velocity. In silicon, this occurs at a field strength of 2 V μm^{-1} which implies a bias voltage of 80 V for a 40-μm-thick I-layer.

4.3.3 Speed Limitations

If a photodiode is to detect a digital signal, the sum of the rise- and fall-times of the electrical signal must be less than the interval between optical pulses. If we cannot satisfy this condition, then inter-symbol interference, *ISI*, will occur. Three main factors limit the photodiode response time: carrier diffusion time from the p^+ and n^+ regions, carrier transit time across the I-layer and the junction capacitance interacting with any external load resistance.

As regards the photodiode capacitance, its effect can be reduced by boot-strapping. (This is dealt with in Chap. 6.) What can be of importance is the transit time response which can ultimately lead to a limit on bandwidth.

As we have previously seen, photon absorption can occur in the p^+ region of the photodiode. As the optical power falls exponentially as we move through the diode, this region will produce a considerable number of EHPs. If the EHPs are produced within a diffusion length of the depletion region, the electrons will diffuse into the I-layer where they will ultimately contribute to current flow. In principle, as soon as an EHP is produced in the p^+ region, a current will flow because of the production of a hole. However, the duration of the current pulse will be longer than the optical pulse because the electrons have to cross the I-layer. If the thickness of the p + layer is greater than a diffusion length, the maximum transit time of the electrons will be the sum of the electron lifetime in the p + region ($\approx$10 ns) and the transit time across the I-layer ($\approx$ 0.1 ns). So, in order to produce a fast device, we must ensure that the electrons do not spend a long time in the p + layer, that is, we must use a p + region that is considerably less than a diffusion length.

Let us now consider EHP generation in the I-layer. If photon absorption occurs near the p + region, the electrons are accelerated across the I-layer by the electric field. If we consider a 40-μm-thick I-layer, then $E = 0.3$ V μm^{-1} at a bias of 12 V. The mobility of electrons in intrinsic silicon is 1350 cm^2 V^{-1} s^{-1}, and so the electron velocity is 4×10^4 m s^{-1} giving a transit time of approximately 1 ns. If we can operate with a sufficiently high E field, the electrons will reach their saturation velocity of 10^5 m s^{-1} resulting in a maximum transit time of 0.4 ns (considerably lower than the diffusion time of electrons in the p + region).

If EHP generation occurs close to the n + region, it is the holes that have to traverse the I-layer. The saturation velocity of holes in silicon is 0.5×10^5 m s^{-1}, and this gives a transit time of 0.8 ns. However, we should remember that the optical power is quite low close to the n + region, and so there will not be many EHPs produced. Thus, in general, we can neglect the effects of EHP generation in this

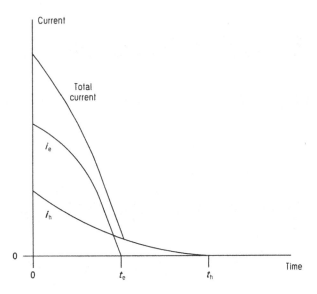

Fig. 4.7 Theoretical impulse response of a silicon PIN photodiode

region. It is interesting to note that if the diode is illuminated from the n + side, it will be the holes that have to traverse the I-layer. As holes are slower than electrons, the falling edge of the current pulse will be much slower than if electrons were the carrier.

So, we have seen that the current pulse produced by an optical signal will have a fast rise-time, with a fall-time governed by diffusion effects and hole transit across the I-layer. Figure 4.7 shows the theoretical impulse response of a silicon PIN photodiode. Also shown are the individual contributions due to electron and hole propagation. As can be seen, it is the diffusion of electrons that causes the poor fall-time.

So far we have ignored the effects of the depletion layer capacitance. If we initially neglect the package capacitance, the diode capacitance will be given by

$$C = \varepsilon_o \varepsilon_r \frac{A}{W} \qquad (4.15)$$

where A is the cross-sectional area of the diode and W is the depletion region thickness. By itself the diode capacitance presents few problems. However, when connected to an external load, the RC time constant may be sufficient to limit the maximum frequency of operation. Total diode capacitances (including package capacitance) range from less than 0.5 pF for high-speed detectors to 150 pF for low-speed, large area detectors.

Example

A Si PIN photodiode has a 0.5 μm p⁺ layer, a 40 μm n⁻ layer and a 10 μm n + layer. An anti-reflection coating is used that results in 10 per cent reflection at 850 nm, and the diode has a circular cross-section of diameter 50 μm. Determine the quantum efficiency and examine the limits on detector speed. Assume that the diode is operated with a reverse bias of 20 V.

The diode is 10 per cent reflective, and so 90 per cent of the incident light reaches the silicon. The dimensions of the p⁺ and n⁺ layers are such that any carriers generated in these regions will diffuse into the depletion region. In passing through the diode, 90 per cent of the light is absorbed. Thus the overall quantum efficiency of the diode is 86 per cent.

As regards the speed of the detector, the p + region is thin enough for us to neglect any diffusion of carriers to the I-layer. So, the main limit on the response time is the transit time of carriers across the I-layer. In order to find this transit time, we need to find the electric field strength across the I-layer. This field strength is

$$E = \frac{20}{40 \times 10^{-6}} = 0.5 \text{ V}/\mu\text{m}$$

If we take an electron mobility of 1350 cm^2 V^{-1} s^{-1}, we get an electron velocity of 6.75×10^4 m s^{-1} which results in an electron transit time of approximately 600 ps. By following a similar analysis for the holes (500 cm^2 V^{-1} s^{-1} mobility), we get a hole transit time of 1.6 ns. As we have already noted, the optical power is low close to the n + region and so we can usually neglect the effects of hole transit time.

As regards the diode capacitance, the cross-sectional area of the diode is approximately 2×10^3 μm^2 which results in a depletion capacitance of 5 fF. This value is low enough for any package capacitance to dominate.

4.3.4 Photodiode Circuit Model

Figure 4.8 shows an equivalent circuit for a PIN photodiode, which is connected to an external load feeding an amplifier. In this diagram, the photoconductive current has been modelled as a current source, I_s, whose magnitude depends on the incident optical power. The constant current source, I_d, models the dark current, that is, the leakage current and any photoconductive current due to background radiation. The shunt resistance, R_j, represents the slope of the reverse bias characteristic, and the series resistance, R_s, is that of the bulk semiconductor and the contact resistance. The *total* diode capacitance, C_d, models the depletion and diffusion capacitances. The load resistor, R_L, shunts this capacitance, and it is this time constant that usually limits the speed of response.

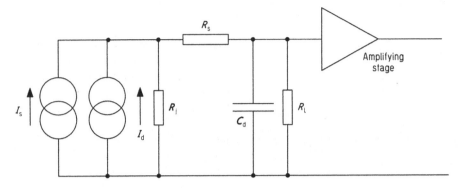

Fig. 4.8 A circuit model for a typical photodiode

In general, we can ignore R_j and R_s and so the bandwidth of the detector is given by

$$f = \frac{1}{2\pi R_L C_d} \qquad (4.16)$$

We should also note that any following amplifier will have a time constant, and this may prove to be the limiting factor.

4.3.5 Long-Wavelength PIN Photodiodes

At long wavelengths, >1 μm, silicon becomes transparent. Thus detectors for 1.3 and 1.55 μm wavelengths must be made out of low band-gap materials. Germanium has a band-gap of 0.67 eV, corresponding to a cut-off wavelength of 1.85 μm, and so would appear to be a suitable material. However, the low band-gap means that Ge photodiodes exhibit a high leakage current (>100 nA). As we will see later, the dark current is an additional source of noise, and so Ge PIN photodiodes are rarely used in long-haul routes.

When we discussed light sources, we saw that InGaAsP emits light in the band 1.0–1.7 μm. Thus detectors made of a similar material should respond to 1.3 or 1.55 μm light. In practice, we can use an InGaAs alloy, where the proportions of In and Ga alter the band-gap. Thus the diode can be tailored to respond to light of a specific wavelength. As an example, a diode fabricated out of $In_{0.53}Ga_{0.47}As$ has a band-gap of 0.47 eV which gives a cut-off wavelength of 1.65 μm. The dark current of these devices is usually about 10 nA.

The absorption coefficient of InGaAs at 1.3 μm is about 5×10^5 m^{-1} which results in a penetration depth of around 2 μm. Therefore the dimensions of a long-wavelength PIN detector are much smaller than that of a Si photodiode, leading to a

Fig. 4.9 Cross-section through a typical long-wavelength PIN photodiode

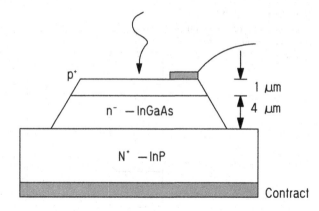

better frequency response. Figure 4.9 shows the structure of a typical InGaAs photodiode.

The quantum efficiency of this particular device is quite low, typically 0.4, because the p + layer absorbs 40 per cent of the incident power. However, the InP substrate is transparent to light of wavelength greater than 0.92 μm. Thus illumination from the rear of the device will increase the quantum efficiency to about 90 per cent. Such a device is known as a *rear-entry* or *substrate-entry* photodiode.

The I-layer is usually doped to a level of 10^{21} m^{-3}, and this, together with an ε_r of 14, gives a punch-through voltage of 10 V. This results in an E field of 2.5 V μm^{-1} which is well above the 1 V μm^{-1} required for the carriers to reach their saturation velocity of about 1×10^5 m s^{-1}. So the transit time across the I-layer is in the region of 40 ps, and, as there is little absorption in the p + layer, the device is inherently very fast.

The junction capacitance of a typical 50-μm-diameter device is 60 fF, which is considerably less than that of a Si diode of the same dimensions. However, any package capacitance may cause the capacitance to rise to 0.8 pF or more. Thus for high-speed operation, hybrid thick-film receivers use unpackaged photodiodes.

4.4 Avalanche Photodiodes (APDs)

When a p-n junction diode has a high reverse bias applied to it, breakdown can occur by two separate mechanisms: direct ionisation of the lattice atoms, *zener breakdown*, and high-velocity carriers causing impact ionisation of the lattice atoms, *avalanche breakdown*. APDs use the latter form of breakdown.

Figure 4.10 shows the schematic structure of an APD. By virtue of the doping concentration and physical construction of the n + p junction, the E field is high enough to cause impact ionisation. Under normal operating bias, the I-layer (the p-region) is completely depleted. This is known as the *reach-through* condition, and so APDs are sometimes known as *reach-through APDs* or *RAPDs*.

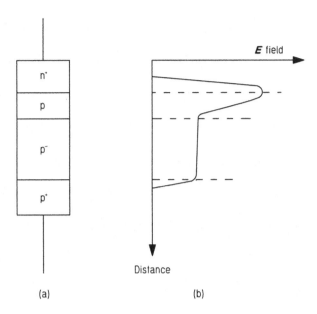

Fig. 4.10 (**a**) APD schematic and (**b**) variation of electric field intensity across the diode

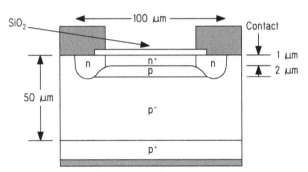

Fig. 4.11 Cross-section through a typical silicon APD

Like the PIN photodiode, light absorption in APDs is most efficient in the I-layer. In this region, the E field separates the carriers and the electrons drift into the avalanche region where carrier multiplication occurs. We should note however that an APD biased close to breakdown could break down owing to the reverse leakage current. Thus APDs are usually biased just below breakdown, with the bias voltage being tightly controlled.

4.4.1 APD Structures

Figure 4.11 shows the cross-section of a typical Si APD. In order to minimise photon absorption in the n + p region, the n + and p layers are made very thin. In practice, these layers have doping concentrations of around 10^{24} and 10^{21} m^{-3}, and the

Fig. 4.12 Cross-section
through a typical long-
wavelength
heterojunction APD

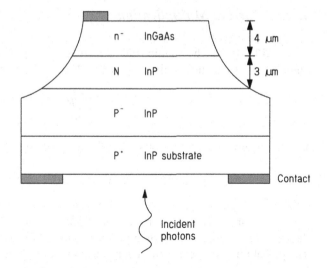

p + and p- layers have concentrations of 10^{24} and 10^{20} m^{-3}, respectively. These
parameters, together with the device dimensions, result in a reach-through voltage of
approximately 40 V and an avalanche breakdown E field value of typically
18 V μm^{-1}. (The reverse breakdown voltage for a typical device lies in the range
200–300 V.) An n-type guard ring serves to increase the peripheral breakdown
voltage, causing the n + p junction to breakdown before the pn junction.

For operation at 1.3 and 1.55 μm wavelengths, we can use germanium APDs.
However, as we have already noted, these diodes exhibit a high dark current, giving
a noisy detection process. In spite of their drawbacks, several different Ge APD
structures are being investigated, including SiGe compound APDs, and such devices
may find applications in the future.

Like long-wavelength PIN photodiodes, APDs can be made out of InP/InGaAs,
and Fig. 4.12 shows a typical structure. In this particular design, the InGaAs layer
absorbs the light, and, because InP is transparent to long-wavelength light, the
device can be either front or rear illuminated. The E field in the fully depleted region
causes separation of the photo-generated carriers. However, because of the n-N
heterojunction, only holes cause breakdown in the N-type region.

Such APDs usually operate with a sufficiently high E field in the absorbing
region, >1 Vμm^{-1}, to accelerate the carriers to their saturation velocity and a field
strength in the N-type large enough to cause breakdown, >20 V μm^{-1}. In practical
devices, the operating fields are typically 15 V μm^{-1} and 45 V μm^{-1}, and so these
conditions are satisfied. The bias voltage at which these fields occur is usually
around 50 V.

4.4.2 Current Multiplication

In an APD, avalanche multiplication increases the primary current, that is, the unmultiplied photocurrent given by (4.12). Thus we can write the responsivity as

$$R_o = \frac{M q \eta \lambda_o}{hc} \tag{4.17}$$

where M is the multiplication factor. It therefore follows that M is given by

$$M = \frac{I_m}{I_s} \tag{4.18}$$

where I_m is the average total multiplied diode current. In order for M to be large, there must be a large number of impact ionisation collisions in the avalanche region. The probability that a carrier will generate an electron-hole pair in a unit distance is known as the *ionisation coefficient* (α_e for electrons and α_h for holes). Obviously, M is highly dependent on these coefficients which, in turn, depend upon the E field and the device structure. After a straightforward analysis, it can be shown that M is given by

$$M = \frac{1 - k}{\exp\left[-(1 - k)\alpha_e W\right] - k} \tag{4.19}$$

where k is α_e/α_h and W is the width of the avalanche region. So, a large M requires a low value of k. In silicon, k ranges from 0.1 to 0.01, and this leads to values of M ranging from 100 to 1000. However, in germanium and III-V materials, k ranges from 0.3 to 1 and, in practice, it is difficult to fabricate and control devices with gains above 15.

As expected, M is highly dependent on the bias voltage. An empirical relationship which shows this dependency is

$$M = \frac{1}{1 - (V/V_{br})^n} \tag{4.20}$$

where V_{br} is the device breakdown voltage and n is an empirical constant, <1. Now, n and V_{br} are dependent on temperature as shown by

$$V_{br}(T) = V_{br}(T_o) + a(T - T_o)$$

and

$$n(T) = n(T_o) + b(T - T_o) \tag{4.21}$$

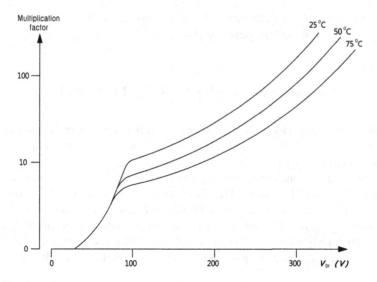

Fig. 4.13 Theoretical variation of multiplication factor, M, with reverse bias voltage, V_{br}, for three different temperatures. Unity gain has been taken at $V_{br} = 30$ V

where a and b are empirical constants, <1. So, as Fig. 4.13 shows, M depends on both the bias voltage and the temperature.

4.4.3 Speed Limitations

Several factors will limit the speed of response of an APD: the RC time constant of the detector circuitry; the drift time of carriers to the avalanche region, t_d; the time taken to achieve avalanche breakdown, t_a; and the time taken to sweep the avalanche produced carriers through the diode, t_s. Of these four factors, t_a and t_s represent delays which are additional to those experienced with PIN photodiodes.

A full analysis of the APD response times reveals that the intrinsic time constant, τ, for an APD with $k \ll 1$ is given by

$$\tau = t_d + t_a + t_s$$
$$= \frac{W_i}{v_{se}} + \frac{MkW_a}{v_{se}} + \frac{1}{v_{sh}}(W_a + W_i) \tag{4.22}$$

where W_i and W_a are the widths of the intrinsic and avalanche regions, respectively, and v_{se} and v_{sh} are the saturation velocities of electrons and holes, respectively. As can be seen, a fast diode requires $k \ll 1$. Silicon has $k = 0.05$, and so this is a very popular material. In general, APDs have a slower response time than an equivalent PIN photodiode, and so gain has been traded for a reduction in bandwidth. It should be noted that the RC time constant of the diode capacitance and the external load is likely to limit the overall frequency response.

4.5 Metal Semiconductor Metal (MSM) Photodiodes

Schottky diodes are widely used in microwave mixers where their fast switching speed is used to good effect. In optical communications, the Schottky diode takes the form shown in Fig. 4.14.

In this diode, a junction is formed between the semiconductor material (Si or GaAs) and two metal contacts. This forms two Schottky diodes. The form of the electrodes is worthy of note. The comb-like structure is known as interdigitated electrodes. This generates a low contact capacitance contact as the electrode capacitance is effectively made up a series of capacitors in series. This means the diode can have a very fast response (<5 ps) making them ideal for use at high bit rates.

4.6 Photodiode Noise

The signal at the end of any optical link is often highly attenuated, and so any receiver noise should be as small as possible. The minimum signal-to-noise ratio, S/N, required for satisfactory detection is often specified for a particular application. This is dealt with in greater detail in the next chapter; however, the S/N can be written as

Fig. 4.14 Schematic of a MSM photodiode

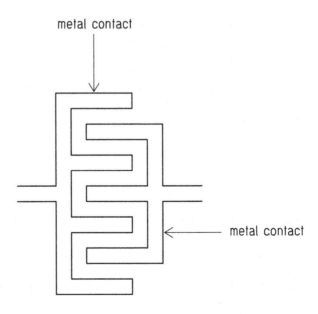

$$\frac{S}{N} = \frac{<I_s^2>}{<i_n^2>_{pd} + <i_n^2>_c} \tag{4.23}$$

where $<I_s^2>$ is the mean square, *m.s.*, value of the photodiode signal current, $<i_n^2>_{pd}$ is the m.s. value of the photodiode noise and $<i_n^2>_c$ is the m.s. value of the following amplifier noise when referred to the detector terminals. Even if the following amplifier is noiseless, there is still some photodiode noise, and it is this noise source that concerns us here.

There are three main components to the photodiode noise: *quantum noise*, $<i_n^2>_Q$, due to quanta of light-generating packets of electron-hole pairs; thermally generated dark current, $<i_n^2>_{DB}$, occurring in the photodiode bulk material; and surface leakage current, $<i_n^2>_{DS}$. (There is an extra noise component due to the ambient light level causing additional dark current; however, this is more likely in free-space links and careful shielding of the detector can reduce this to a minimum.) Thus we can write $<i_n^2>_{pd}$ as

$$<i_n^2>_{pd} = <i_n^2>_Q + <i_n^2>_{DB} + <i_n^2>_{DS} \tag{4.24}$$

4.6.1 PIN Photodiode Noise

No current multiplication occurs in a PIN detector, and so, with a receiver noise equivalent bandwidth of B_{eq} (Appendix 1), we can write the detector *S/N* as

$$\frac{S}{N} = \frac{<I_s^2>}{<i_n^2>_Q + 2qI_{DB}B_{eq} + 2qI_{DS}B_{eq}} \tag{4.25}$$

where I_{DB} and I_{DS} are the bulk and surface leakage currents, respectively. However, if the leakage currents are negligible

$$\frac{S}{N} = \frac{<I_s^2>}{<i_n^2>_Q} \tag{4.26}$$

With this condition, we must take account of the quantum nature of light and so the S/N defined by (4.26) is known as the *quantum limit*. We can determine $<i_n^2>_Q$ from a knowledge of the photon statistics.

Photons arrive at the detector at random intervals, but with a constant *average* rate. So, in a certain time interval, we can expect to receive an average of m photons but, because of the random arrival of photons, we actually receive n photons. The photon arrival follows a Poisson probability distribution and so the probability that the resultant number of detected photons is n, with an expected number of m, is

$$p(n) = \text{Pos}[n, m] = \frac{m^n e^{-m}}{n!} \qquad (4.27)$$

If we take a quantum efficiency of unity, the number of EHPs is n. In a digital system, a decision must be made as to whether a 1 or a 0 was sent; however, noise will corrupt the signal levels so that a 1 signal occasionally turns into a 0 and a 0 turns into a 1. In an ideal receiver, the detection of a single EHP results in a logic 1, while the absence of any signal current results in a logic 0. As the quantum noise is dependent on the *presence* of an optical signal, it will only corrupt logic 1 signals (assuming no dark current and no other noise sources). So, the condition for a logic 1 detection error is that m photons are received, but $n = 0$ photons are detected. If we assume that the probability of sending a logic 1 is the same as for a logic 0, that is, they are *equiprobable*, we can write the probability of an error, P_e, as

$$P_e = \frac{1}{2}\left(P(0|1) + P(1|0)\right)$$

$$= \frac{1}{2}\left(\frac{m^0 e^{-m}}{0!} + 0\right) \qquad (4.28)$$

So, for a typical error rate of 1 bit in 10^9, we require an average of $m = 21$ EHPs. These carriers are generated by $21/\eta$ photons arriving in a bit-time $1/B$, where B is the data rate and η is the quantum efficiency. For equiprobable 1s and 0s, the *mean* optical power required, P, is

$$P = \frac{1}{2}\frac{21}{\eta}\frac{hc}{\lambda_o}B \qquad (4.29)$$

Example
Determine the quantum limit for a PIN photodiode, with unity quantum efficiency, that detects 1 Gbit/s data at a wavelength of 1550 nm. Assume an error rate of 1 bit in 10^9.
    If we use (4.29), we find that the mean optical power for the specified error rate is
    $P = 1.35$ nW
        $= -58.7$ dBm

(continued)

This represents a very high sensitivity. However, this result ignores the noise from the photodiode dark current and following amplifier stages. In practice, these effects will tend to limit the receiver sensitivity. In spite of this, coherent detection systems (which we examine in the final chapter) can achieve sensitivities better than the direct detection quantum limit we are considering here.

Before we comment on APD noise, we should note that the spectral density of the quantum noise is simply the shot noise expression, given by

$$< i_n^2 >_Q = 2q < I_s > \quad \mathrm{A^2/Hz} \tag{4.30}$$

where $<I_s>$ is the mean signal current. (This result arises from the statistics of the Poisson process.) Therefore we can write the quantum limited S/N as

$$\frac{S}{N} = \frac{< I_s^2 >}{2q < I_s^2 > B_{eq}} \tag{4.31}$$

4.6.2 APD Noise

In an APD, the avalanche gain multiplies the *primary current*. (We define the primary current as that produced by a unity gain photodiode.) Since the gain is statistically variant, that is, not all of the photo-generated carriers undergo the same multiplication, we define the *average* gain as M. As we have seen, the Poisson distribution describes photon arrival and hence the signal current. So, we can find the APD signal current by performing a convolution type process between the Poisson distribution of the primary current and the avalanche gain distribution. The resulting expression is very complicated, and so it is common practice to approximate the APD current to a Gaussian distribution with a mean value of $<I_s>M$ and a noise current spectral density of $2q<I_s>M^2 F(M)$. The term $F(M)$ is known as the *excess noise factor*, and we include it to account for the random fluctuations of the APD gain about the mean. We can approximate $F(M)$ by

$$F(M) = M^x \tag{4.32}$$

where x is an empirical constant which is less than unity. From our earlier discussion of avalanche multiplication, it should be apparent that $F(M)$ depends on the value of k and the type of carrier undergoing multiplication. Detailed analysis (McIntyre [1]) shows that $F(M)$ can be approximated by

$$F_e(M) = kM_e + \frac{(1-k)}{M_e}(2M_e - 1) \tag{4.33}$$

for electron avalanche and

$$F_h(M) = \frac{M_h}{k} + \frac{(1-1/k)}{M_h}(2M_h - 1) \tag{4.34}$$

for hole avalanche. These equations clearly show the need to fabricate devices out of materials with low values of k.

Example
A Si APD has a gain of 100 and $k = 0.02$. Determine the excess noise factor for electrons and compare it with that obtained from a Ge APD with $M = 20$ and $k = 0.5$.
By applying (4.33), we find that $F_e(M)$ for electrons in the silicon APD is

$$F_e(M) = kM_e + \frac{(1-k)}{M_e}(2M_e - 1)$$

$$= 0.02 \times 100 + \frac{(1-0.02)}{100}(2 \times 100 - 1)$$

$$\approx 4$$

For the germanium diode however, $F_e(M)$ is

$$F_e(M) = 0.5 \times 20 + \frac{(1-0.5)}{20}(2 \times 20 - 1)$$

$$\approx 11$$

Thus we can see that as well as having higher gains than Ge APDs, Si APDs have a lower excess noise factor.

As regards the S/N for an APD, we have to take account of signal multiplication and the noise terms due to surface and bulk leakage currents. Thus we can write the S/N as

$$\frac{S}{N} = \frac{<I_s^2>M^2}{2q<I_s>M^2F(M)B_{eq} + 2qI_{DB}M^2F(M)B_{eq} + 2qI_{DS}B_{eq} + <i_n^2>_c} \tag{4.35}$$

If we ignore the leakage currents and assume a noiseless receiver, we can write (4.35) as

$$\frac{S}{N} = \frac{<I_s^2> M^2}{2q <I_s> M^2 F(M) B_{eq}} \tag{4.36}$$

Comparison with the PIN Eq. (4.31) reveals that, because of the excess noise factor, an APD receiver cannot approach the quantum limit. Indeed, at high levels of received signal power, the use of an APD could be a disadvantage because of the signal-dependent shot noise. However, as we will presently see, we can use an APD to increase the signal-to-noise ratio of an ordinarily noisy optical receiver. If the noise from the following amplifier stage is greater than that of the detector noise, the S/N approximates to

$$\frac{S}{N} = \frac{<I_s^2> M^2}{<i_n^2>_c} \tag{4.37}$$

Thus, because of the M^2 term, the S/N for an APD receiver can be greater than that for a PIN receiver. As the following example demonstrates, in most APD receivers, the sensitivity advantage reduces because we cannot ignore the detector noise. (Section 5.4 in the next chapter also deals with this problem.)

Example

Two receivers are available for use in an optical link: one has a total input equivalent noise current of 10^{-15} A^2, while the other has an $<i_n^2>$, of 10^{-18} A^2. Both receivers have a noise equivalent bandwidth of 27 MHz. Determine the S/N of the receivers if the mean received signal power is 100 nW. Assume a diode responsivity of 0.5 A/W, M = 100, F(M) = 4, $I_{DB} = I_{DS} = 10$ nA.

As the diode responsivity is 0.5 A/W, the primary signal current is

$<I_s> = 0.5 \times 100$ nA

$= 50$ nA

In order to find S/N, we can use (4.35) to give, for the noisy receiver,

$$\frac{S}{N} = \frac{<I_s^2> M^2}{2q <I_s> M^2 F(M) B_{eq} + 2q I_{DB} M^2 F(M) B_{eq} + 2q I_{DS} B_{eq} + <i_n^2>_c}$$

$$= \frac{(50 \times 10^{-9})^2 \times 100^2}{1.73 \times 10^{-14} + 3.5 \times 10^{-15} + 8.6 \times 10^{-20} + 1 \times 10^{-15}}$$

$= 1.15 \times 10^3$

$= 30.60$ dB

For the less noisy receiver, we get an S/N of 30.80 dB. Thus we can see that there is negligible difference between the two receivers. This is because the dominant noise source is the first term in the denominator of (4.35) – the signal-dependent noise.

(continued)

It is instructive to compare these sensitivities to those that would be achieved if a PIN photodiode, with the same basic parameters, were used. In this instance, the S/N for the noisy receiver is

$$\frac{S}{N} = \frac{\left(50 \times 10^{-9}\right)^2}{4.32 \times 10^{-19} + 8.6 \times 10^{-20} + 8.6 \times 10^{-20} + 1 \times 10^{-15}}$$
$$= 2.5$$
$$= 4.00 \text{ dB}$$

while the *S/N* for the less noisy receiver is 31.95 dB. So, a considerable sensitivity advantage results from using the less noisy receiver.

We can see from these figures that the signal-dependent shot noise associated with an APD tends to mask the effects of receiver noise. However, in a PIN photodiode receiver, a considerable sensitivity advantage results from using a low-noise receiver. We can also see that the PIN detector and a low-noise receiver offer a slightly better sensitivity than the equivalent APD receiver. If we increase the mean signal power to 1 μW, the PIN detector gives a 5 dB advantage over the APD detector if the low-noise receiver is used. This is due to the effect of the excess noise factor which means that an APD cannot reach the quantum limit, whereas a PIN detector can. So, in general, we can conclude that we can use an APD to increase the sensitivity of a noisy receiver. However, a PIN detector gives a better sensitivity when detecting high power levels. The exact advantage depends on the bulk leakage current, the magnification facto, and the excess noise factor.

Problems

1. A p$^+$n diode is formed from InGaAs which is doped with $N_a = 10^{25}$ m^{-3} and $N_d = 10^{22}$ m^{-3}. Determine the zero-bias barrier voltage if the density of thermally generated carriers is 6.3×10^{17} m^{-3}. Determine the width of the depletion region at a reverse bias of 10 V. Assume $\varepsilon_r = 13.1$.

[1.8 V, 1.3 μm]

2. A PIN photodiode uses an InGaAs, 4-μm-thick intrinsic region which is doped to a level of 1021 m^{-3}. Determine the punch-through voltage.

[9.2 V]

3. For the PIN diode of Q2, determine the amount of light not absorbed by the I-layer given that the absorption coefficient is 1.3×105 cm^{-1}.

[0]

4. Determine the capacitance of the diode of Q2 if it has a circular cross-section with a diameter of 50 μm.

[57 fF]

5. Determine the excess noise factor of a InGaAs APD for use at 1550 nm if the APD gain is 10 and k is 0.8.

[8.4]

6. Determine the S/N for a receiver using the APD of Q5. The noise of the receiver is 1×10^{-15} A^2, the noise equivalent bandwidth is 1 GHz, the dark currents are 10 nA, and the signal current is 3 μA.

[30.5 dB]

Recommended Reading

1. McIntyre RJ (1966) Multiplication noise in uniform avalanche diodes. IEEE Trans Electron Devices ED-13:164–168
2. McIntyre RJ, Conradi J (1972) The distribution of gains in uniformly multiplying avalanche photodiodes. IEEE Trans Electron Devices ED-19:713–718
3. Stillman GE et al (1983) InGaAsP photodiodes. IEEE Trans Electron Devices ED-30:364–381
4. Kressel H (ed) (1980) Semiconductor devices for optical communications, Topics in applied physics, vol 39. Springer, New York

Chapter 5
Introduction to Receiver Design

The basic structure of an optical receiver (Fig. 5.1) is similar to that of a direct detection r.f. receiver: a low-noise preamplifier, the *front-end,* feeds further amplification stages, the *post-amplifier,* before filtering. An important point to note is that the pre- and post-amplifiers are usually non-saturating. (If the amplifiers did saturate, charge storage in the transistors would tend to limit the maximum detected data rate.) Because of this, we can use the same pre- and post-amplifier combination to detect analogue or digital signals. The difference between the two receivers arises from the way they process the signals after amplification. As digital optical communications systems are quite common, most of the work presented is devoted to a performance analysis of digital receivers. However, analogue systems are used to transmit composite video and signals from optical fibre sensors, and so we will consider analogue receiver performance towards the end of this chapter. Although preamplifier design is dealt with in the next chapter, we must make certain assumptions regarding its performance: the bandwidth must be large enough so as not to distort the received signal significantly; and its gain function must be high enough so that we can neglect any noise from the following stages. As we shall see later, the requirement to minimise the noise implies restricting the receiver bandwidth. However, a low bandwidth results in considerable inter-symbol interference, ISI, and so the receiver bandwidth is a compromise between minimising the noise and ISI.

5.1 Fundamentals of Noise Performance

In order to examine the noise performance of an optical receiver, and hence determine its sensitivity, we shall consider the receiver as a linear channel, with the a.c. equivalent circuit shown in Fig. 5.2.

An ideal current source, shunted by the detector capacitance, C_d, models the photodetector, which feeds the parallel combination of R_{in} and C_{in} modelling the input impedance of the preamplifier. The pre- and post-amplifiers are modelled as a

© Springer Nature Switzerland AG 2020
M. Sibley, *Optical Communications*, https://doi.org/10.1007/978-3-030-34359-0_5

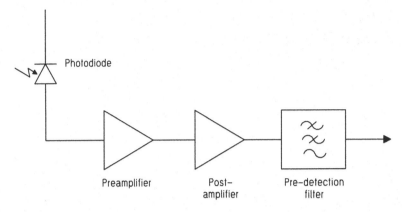

Fig. 5.1 The basic structure of an optical receiver

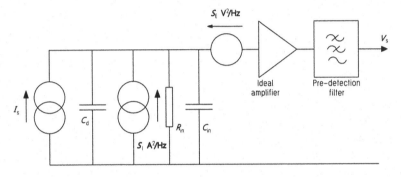

Fig. 5.2 A.c. equivalent circuit of an optical receiver

single voltage amplifier, with transfer function $A(\omega)$, the output of which feeds the pre-detection filter. If we initially neglect the photodiode noise, then the only noise in the receiver will be due to the preamplifier. A shunt noise generator S_I, with units of A^2/Hz, models the noise current due to the preamplifier first stage and the photodiode load resistor. The series noise generator S_E with units of V^2/Hz models the preamplifier series noise sources. (The reason for the inclusion of this generator will become apparent when we consider preamplifier design in the next chapter.)

In order to determine the signal-to-noise ratio, S/N, at the output of the pre-detection filter, we need to find the receiver transfer function. Because the input signal is a current, I_s, and the output is a voltage, V_s, the transfer function is a *transimpedance*, $Z_T(\omega)$, given by

$$Z_T(\omega) = \frac{V_s}{I_s} \tag{5.1}$$

From Fig. 5.2, we see that

$$V_s = I_s Z_{in} A(\omega) H_f(\omega) \tag{5.2}$$

where Z_{in} is the total input impedance, that is, the parallel combination of R_{in} and the *total* input capacitance $(C_d + C_{in})$ and $H_f(\omega)$ is the pre-detection filter transfer function. Thus we can express $Z_T(\omega)$ as

$$Z_T(\omega) = Z_{in} A(\omega) H_f(\omega) \tag{5.3}$$

If we now turn our attention to the noise sources, we can see that the series noise generator produces a m.s. *input* noise current of

$$\frac{S_E}{Z_{in}^2} \quad \text{or} \quad S_E Y_{in}^2 \ A^2/Hz$$

If we assume that the two noise sources are independent of each other, that is, *uncorrelated*, then the total equivalent input noise current spectral density, $S_{eq}(f)$, will be given by

$$S_{eq}(f) = S_I + S_E Y_{in}^2$$

Noting that

$$Y_{in} = \frac{1}{R_{in}} + j\omega C_T$$

where C_T is $C_d + C_{in}$, we can write $S_{eq}(f)$ as

$$S_{eq}(f) = S_I + S_E\left(\frac{1}{R_{in}^2} + (2\pi C_T)^2 f^2\right) \tag{5.4}$$

Thus we can see that the equivalent input noise current spectral density consists of two terms: a frequency-independent term and a term that varies according to f^2. (We will return to this point in the next chapter.)

Now, with $S_{eq}(f)$ given by (5.4), we can write the total m.s. output noise voltage, $<n^2>_T$ as

$$\begin{aligned} <n^2>_T &= \int_0^\infty S_{eq}(f)[Z_T(\omega)]^2 df \\ &= \left(S_I + \frac{S_E}{R_{in}^2}\right)\int_0^\infty [Z_T(\omega)]^2 df + S_E(2\pi C_T)^2 \int_0^\infty [Z_T(\omega)]^2 f^2 df \end{aligned} \tag{5.5}$$

Except for $Z_T(\omega)$, which depends on the filter characteristic, we can find all the parameters in (5.5) from a knowledge of the preamplifier design, dealt with in the next chapter. In the following section, we shall examine $Z_T(\omega)$ in greater detail. In particular, we will determine the frequency response of a digital receiver which results in the minimum output noise, while retaining an acceptable degree of ISI.

5.2 Digital Receiver Noise

In order to determine the integrals in (5.5), we redefine $Z_T(\omega)$ as

$$Z_T(\omega) = R_T H_T(\omega) \tag{5.6}$$

where R_T is the low-frequency transimpedance and $H_T(\omega)$ represents the frequency dependence of $Z_T(\omega)$. If $H_p(\omega)$ is the Fourier transform, FT, of the received pulse, $h_p(t)$, and $H_{out}(\omega)$ is the FT of the pulse at the output of the filter, $h_{out}(t)$, then we can express $Z_T(\omega)$ as

$$Z_T(\omega) = R_T H_T(\omega) = \frac{H_{out}(\omega)}{H_p(\omega)}$$

If we now normalise the output pulse shape, that is, remove the dependency on R_T, we can write

$$H_T(\omega) = \frac{H_{out}(\omega)}{H_p(\omega)} \tag{5.7}$$

The FTs used in (5.7) depend on the bit-time of the pulses, T seconds. In order to remove this dependency, we use a normalised, dimensionless frequency variable, y, defined by

$$y = \frac{f}{B} = \frac{\omega}{2\pi B} = \frac{\omega T}{2\pi} \tag{5.8}$$

where B is the bit rate. We can now define two new functions

$$H_p'(y) = \frac{1}{T} \times H_p(2\pi y/T) \quad \text{and}$$

$$H_{out}'(y) = \frac{1}{T} \times H_{out}(2\pi y/T)$$

Thus the normalised receiver frequency response becomes

$$H'_T(\omega) = \frac{H'_{out}(\omega)}{H'_p(\omega)} \tag{5.9}$$

Because of the normalisation of $H_T(\omega)$, the integrals in (5.5) will only depend on the relative shapes of the input and output pulses. So, to return to (5.5), we can write

$$<n^2>_T = \left(S_I + \frac{S_E}{R_{in}^2}\right)R_T^2BI_2 + S_E(2\pi C_T)^2R_T^2B^3I_3 \tag{5.10}$$

where

$$I_2 = \int_0^\infty [H'_T(y)]^2 dy \quad \text{and} \quad I_3 = \int_0^\infty [H'_T(y)]^2 y^2 dy$$

(The inclusion of the B and B^3 terms in (5.10) accounts for the bandwidth (bit rate) dependency of the noise. In fact, we can regard BI_2 and B^3I_3 as the noise equivalent bandwidths for the frequency-independent and f^2-dependent noise sources.) Since the signal output voltage and the r.m.s. output noise are both dependent on R_T, we can refer them to the input of the preamplifier. Thus the m.s. equivalent input noise current is given by

$$<i_n^2>_c = \left(S_I + \frac{S_E}{R_{in}^2}\right)BI_2 + (2\pi C_T)^2 S_E B^3 I_3 \tag{5.11}$$

If we know the required S/N, then it is a simple matter to determine the minimum signal current and hence the minimum optical power. This assumes that we know the value of the I_2 and I_3 integrals. As these depend on the shape of the input and output pulses, we must study them in greater detail. Before we consider the input pulse, let us define an output pulse shape that results in low noise and low ISI.

5.2.1   Raised-Cosine Spectrum Pulses

At the output of the pre-detection filter, samples are taken to determine the polarity of the pulses. For minimum error rate, sampling must occur at the point of maximum signal. However, if ISI is present, adjacent pulses will corrupt the sampled pulse amplitude leading to an increase in detection errors. So, we require an output pulse shape that maximises the pulse amplitude at the sampling instant and yet results in zero amplitude at all other sampling points, that is, at multiples of $1/B$ where B is the data rate.

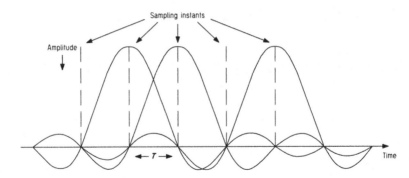

Fig. 5.3 A sequence of sin$x$/x pulses

A sin$x$/x pulse shape will satisfy the ISI requirement. Figure 5.3 shows a sequence of sin$x$/x pulses, and it should be evident that the amplitude of the precursors and tails due to adjacent pulses is zero at the pulse centres. So, the ISI is zero at the sampling instant. A further advantage of these pulses is that the pulse spectrum is identical to the frequency response of an ideal low-pass filter having a bandwidth of $B/2$. As this is the lowest possible bandwidth for a data rate of B, the use of such an output pulse shape results in minimum receiver noise.

There are, however, several difficulties with such an output pulse shape:

1. A receiver transfer function that results in sin$x$/x shape output pulses for a certain input pulse would be very intolerant of any changes in the input pulse shape. Even if the received pulse shape is fixed, variations in component values may cause the bandwidth of the pre-detection filter to reduce, leading to ISI at the sampling instant.
2. It is important to sample at precisely the centre of the pulses, because ISI occurs either side. In practice, the rising edge of the clock varies either side of a mean, a phenomenon known as *clock jitter,* and this results in some ISI. (We can minimise jitter by careful design of the clock extraction circuit; however, some jitter is always present on the recovered clock.)
3. A further disadvantage is that we are considering an ideal sin$x$/x pulse shape. In practice this is impossible to achieve.

From the foregoing, it should be evident that ISI is the major difficulty. The precursors and tails of the sin$x$/x pulses are due to the steep cut-off of the pulse spectrum. So, if we specify a pulse shape with a shallower cut-off spectrum, we can minimise the ISI either side of the sampling instant, so leading to more jitter tolerance. As we will see presently, we can only obtain this advantage at the cost of a reduction in S/N ratio. Let us consider a pulse shape, $h_{\text{out}}(t)$, given by (5.12):

$$h_{\text{out}}(t) = \left(\frac{\sin \pi Bt}{\pi Bt} \right) \times \frac{\cos \pi rBt}{1 - (2rBt)^2} \qquad (5.12)$$

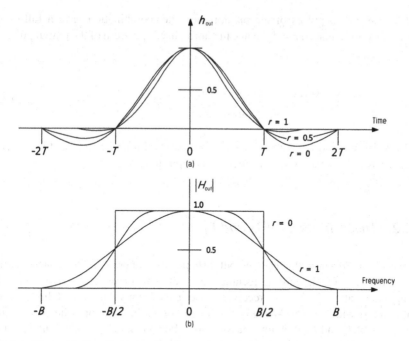

Fig. 5.4 (a) A selection of raised-cosine spectrum pulses and (b) corresponding spectra

As can be seen, a factor that decreases rapidly with time has modified the $\sin x/x$ response of the ideal low-pass filter. Thus, the precursors and tails are considerably reduced, leading to more jitter tolerance and low ISI. The spectrum of these pulses, $H_{out}(f)$, is given by

$$
\begin{aligned}
H_{out}(f) &= 1 && |f| < (1-r)\frac{B}{2} \\
&= \frac{1}{2}[1 + \cos\{(\pi|f| - \pi f_1)/rB\}] && (1-r)\frac{B}{2} < |f| < (1-r)\frac{B}{2} \\
&= 0 \;\; \text{elsewhere}
\end{aligned}
\tag{5.13}
$$

where f_1 is $(1-r)B/2$ and r is known as the *spectrum roll-off factor*. Figure 5.4 shows the normalised pulse shapes and spectra for $r = 0$, 0.5 and 1. As can be seen, the spectra are similar to a cosine that has been shifted up by a d.c. level, and so these pulses are known as *raised-cosine spectrum* pulses. The value of roll-off factor affects both the ISI and the receiver noise: a large roll-off factor gives minimum ISI at the expense of bandwidth, and hence noise; the reverse is true for a low roll-off factor ($r = 0$ yields $\sin x/x$ pulses). In practice, ISI is the more important parameter, and so the output pulses of the preamplifier-filter combination have a full-raised cosine spectrum, that is, $r = 1$. Hence the normalised spectrum of the output pulses is

$$H'_{out}(y) = \frac{1}{2T} \times (1 + \cos \pi y) \quad 0 < |y| < 1$$
$$= 0 \quad \text{elsewhere} \tag{5.14}$$

Provided we know the input pulse shape, we can find the values of the I_2 and I_3 integrals using full-raised cosine spectrum output pulses.

5.2.2   Determination of I_2 and I_3

As we saw in Sect. 2.4, the received pulse shape, $h_p(t)$, depends on the characteristics of the optical link: it may be rectangular, Gaussian, or exponential in form. To complicate matters further, the received pulses may occupy only part of the time slot, that is, short-width pulses. We have to account for all these factors when calculating the values of I_2 and I_3. In a now classic paper, Personick [1] evaluated the integrals for all three different received pulse shapes, and interested readers are referred to the Bibliography for further details. Here we shall consider rectangular shape pulses with normalised FTs:

$$H'_p(y) = \frac{1}{2} \times \frac{\sin \alpha \pi y}{\alpha \pi y} \quad \text{for rectangular pulses} \tag{5.15}$$

The parameter α in (5.15) is the fraction of the time slot occupied by the rectangular pulses. These pulses are shown in Fig. 5.5. If $\alpha = 1$, the pulses fill the whole of the slot, and we have *full-width* or *non-return-to zero*, NRZ, rectangular pulses. Such pulses are normally found in optical links. By using a NRZ input pulse

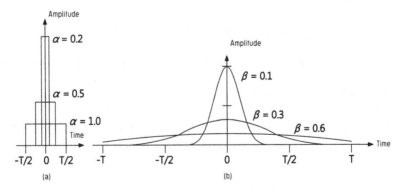

Fig. 5.5 (a) Rectangular and (b) Gaussian shape pulses with various pulse widths

shape, together with raised-cosine spectrum output pulses, we can find the I_2 and I_3 integrals by numerical integration to be $I_2 = 0.564$ and $I_3 = 0.087$.

We can determine the optimum receiver frequency response by dividing the output pulse shape (raised-cosine spectrum pulses) by the input pulse shape (rectangular in shape). For full-width rectangular input pulses, the optimum transfer function is approximated by a single-pole frequency response preamplifier, with a -3 dB cut-off at $B/2$ Hz, feeding a third-order Butterworth filter having a cut-off frequency of $0.7B$ Hz. (A wideband post-amplifier is usually inserted between the preamplifier and the filter.) In applications where the noise performance is not critical, for example, short-haul links, the pre-detection filter is often omitted.

Provided the required S/N is known, we can calculate the receiver noise and hence the receiver sensitivity. In a digital receiver, we can predict the S/N from a knowledge of the decision-making circuitry and the binary signal probabilities. This is the subject of the next section.

5.2.3   Statistical Decision Theory

In a digital receiver, the output of the pre-detection filter consists of a sequence of raised-cosine spectrum pulses in the presence of additive preamplifier noise. The task of any processing circuitry is to determine, with the minimum uncertainty, whether a 1 or a 0 was received. As Fig. 5.6 shows, this is done by a *threshold crossing* device, or comparator, feeding a D-type flip-flop.

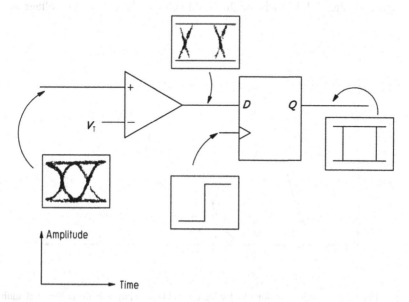

Fig. 5.6 Eye diagrams and schematic diagram of a threshold crossing detector and central decision gate

The circuit operation is best explained by examining the eye diagrams at certain relevant points. (An eye diagram is produced by observing the data stream on an oscilloscope which is triggered by the data clock. Because a complete cycle of the clock corresponds to one bit of data, the eye diagram will show the rising and falling edges of the data, as well as the logic 1 and logic 0 levels.) The eye at the input to the comparator clearly shows the slow rising and falling edges which are due to the limited receiver bandwidth. The effect of the preamplifier noise is to reduce the height and width of the eye, and so the comparator acts to 'clean up' the data. As we shall see presently, the optimum threshold level is mid-way between the logic 1 and 0 levels.

At the output of the comparator, all uncertainty about the level of the pulses has been removed; there is no observable noise at the centre of the eye. However the width of the eye is still affected by noise, and errors could result if sampling occurs close to the cross-over regions. Evidently the point of least uncertainty is the centre of the eye. Thus the clock to the D-type flip-flop is set to latch the data through to the output at the centre of the eye, so-called *central decision detection*. (The precise position of the clock rising edge can be set by using propagation delays through gates and employing various lengths of coaxial cable.) The output of the D-type has no uncertainty associated with it and so a decision has been made, rightly or wrongly, about the received signal.

In order to evaluate the probability of a detection error for a certain *S/N*, we need to examine the noise-corrupted signal, at the input of the comparator, in greater detail. If we assume that 1s and 0s are equiprobable, then we can draw a probability density function plot of the data at the input to the comparator as shown in Fig. 5.7. In this figure, v_{max} and v_{min} represent the *received* signal levels at the output of the pre-detection filter, while V_T is the threshold voltage. So, any signal voltage above

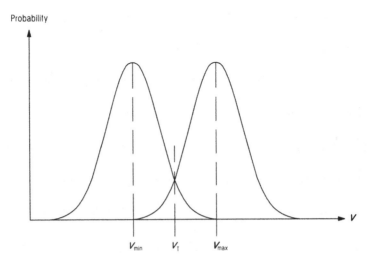

Fig. 5.7 Probability density function plot for logic 1 and logic 0 pulses in the presence of additive Gaussian noise

V_T is received as a logic 1 and any below V_T is a logic 0. In this figure, the area under the logic 0 plot to the right of V_T represents the probability that a zero is received as a one, $P_{e0 \to 1}$. Similarly, the area to the left of V_T is the probability that a logic 1 becomes a logic 0, $P_{e1 \to 0}$. If the noise has a Gaussian distribution

$$P_{e0 \to 1} = \frac{1}{\sqrt{2\pi\sigma_{\mathrm{off}}^2}} \int_{V_T}^{\infty} \exp\left\{-(v - v_{\min})^2/2\sigma_{\mathrm{off}}^2\right\} dv \qquad (5.16)$$

$$P_{e1 \to 0} = \frac{1}{\sqrt{2\pi\sigma_{\mathrm{on}}^2}} \int_{\infty}^{V_T} \exp\left\{-(v_{\max} - v)^2/2\sigma_{\mathrm{on}}^2\right\} dv \qquad (5.17)$$

where σ_{off} and σ_{on} are the r.m.s. noise voltage, at the comparator input, for logic 0 and logic 1 pulses. (The difference between the individual noise voltages accounts for signal-dependent shot noise. Although we are neglecting this noise source for the present, we include these terms for reasons of brevity.) As the probabilities of sending a logic 1 or logic 0 are identical and equal to 1/2, the total error probability, P_e, is

$$P_e = 0.5(P_{e0 \to 1} + P_{e1 \to 0}) \qquad (5.18)$$

If we neglect signal-dependent shot noise, $\sigma_{\mathrm{off}} = \sigma_{\mathrm{on}} = \sigma$. In such circumstances, the optimum threshold voltage lies mid-way between $v_{\max}$ and $v_{\min}$. The reason for this is that if we bias V_T to the left, $P_{e0 \to 1}$ will increase at the expense of $P_{e1 \to 0}$, whereas the opposite is true if we bias V_T slightly to the right. So, with these assumptions, $P_{e0 \to 1} = P_{e1 \to 0}$ and

$$P_e = P_{e0 \to 1}$$
$$= \frac{1}{\sqrt{2\pi\sigma^2}} \int_{V_T}^{\infty} \exp\left\{-(v - v_{\min})^2/2\sigma^2\right\} dv \qquad (5.19)$$

If we change variables by letting

$$x = \frac{v - v_{\min}}{\sigma}$$

we get

$$P_e = \frac{1}{\sqrt{2\pi}} \int_{Q}^{\infty} \exp\left(-x^2/2\right) dx \qquad (5.20)$$

where

$$Q = \frac{V_T - v_{\min}}{\sigma} \qquad (5.21)$$

Since V_T lies mid-way between $v_{\min}$ and $v_{\max}$

$$V_T = \frac{v_{\min} + v_{\max}}{2}$$

and

$$Q = \frac{v_{\max} - v_{\min}}{2\sigma} \qquad (5.22)$$

So, provided we know the signal voltage levels and the r.m.s. noise voltage at the input to the comparator, we can determine the error probability from (5.20). Although we can evaluate this by numerical methods, a more convenient solution is to express it in terms of the widely tabulated *complementary error function, erfc,* as

$$P_e = \frac{1}{2}\mathrm{erfc}\left(Q/\sqrt{2}\right) \qquad (5.23)$$

where

$$\mathrm{erfc}(x) = \frac{2}{\sqrt{\pi}} \int_x^\infty \exp\left(-y^2\right) dy \qquad (5.24)$$

A Q value of 6 results in a bit error rate of 1 in 10^9, that is, $P_e = 1 \times 10^{-9}$ and Fig. 5.8 shows the variation of P_e with Q.

Let us now consider the parameter Q in further detail. As defined by (5.22), all the parameters are directly dependent on the low-frequency transimpedance, R_T, so we can divide throughout by R_T to give

$$Q = \frac{I_{\max} - I_{\min}}{2\sqrt{< i_n^2 >_c}}$$

or

$$\frac{I_{\max} - I_{\min}}{2} = Q\sqrt{< i_n^2 >_c} \qquad (5.25)$$

where $< i_n^2 >_c$ is the m.s. equivalent input noise current as defined by (5.11) and $I_{\max}$ and $I_{\min}$ are the maximum and minimum diode currents resulting from the different light levels. As Fig. 5.9 shows, $I_{\min}$ is not equal to zero. This results from imperfect

Fig. 5.8 Graph of error
probability, P_e, against
signal-to-noise ratio
parameter, Q, for a threshold
crossing detector

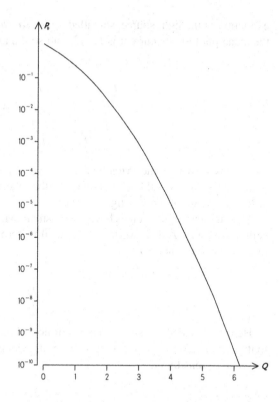

Fig. 5.9 Illustrative of a
non-zero extinction ratio

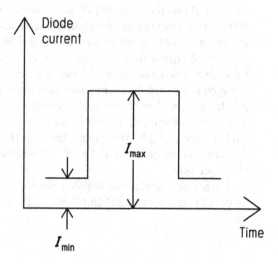

extinction of the light source, so-called *non-zero extinction*. Under these conditions, the mean photodiode current is $I_{max}/2$, and so the receiver sensitivity, P, is

$$
\begin{aligned}
P &= \frac{I_{max}}{2R_o} = \frac{I_{max} - I_{min}}{2R_o} \frac{I_{max}}{I_{max} - I_{min}} \\
&= \frac{Q\sqrt{<i_n^2>_c}}{R_o} \frac{1}{1-\varepsilon}
\end{aligned}
\tag{5.26}
$$

where ε is known as the *extinction ratio* given by I_{min}/I_{max}. (We will return to this point later.) We should note, however, that the error rate is dependent on the *difference* between the two light levels.

Most modern light sources have a very small extinction ratio, and so I_{min} is low in comparison with I_{max}. If we take I_{min} equal to zero, Q becomes the mean signal to r.m.s. noise ratio, and so

$$
P = \frac{Q\sqrt{<i_n^2>_c}}{R_o}
\tag{5.27}
$$

Hence, provided the signal-dependent noise is negligible, we can find the sensitivity from (5.27). In the next section, we will examine the effect of photodiode noise on receiver sensitivity.

Example

An optical receiver detects full-width rectangular pulses at a data rate of 100 Mbit/s. The transfer function of the receiver and pre-detection filter is such that the output pulses have an ideal raised-cosine spectrum. The data consists of equiprobable 1 s and 0 s, and any signal-dependent noise is negligible when compared to the preamplifier noise. The preamplifier noise consists of a frequency-independent term of magnitude 5×10^{-24} A^2/Hz and an f^2 noise term of magnitude 2×10^{-18} V^2/Hz^3. The input impedance of the preamplifier can be taken to be 10 kΩ, and the input capacitance is 3 pF. Determine the sensitivity of the receiver assuming 850 nm wavelength light and a PIN photodiode with a quantum efficiency of 90 per cent.

The receiver detects 100 Mbit/s, full-width rectangular pulse data and so the values of I_2 and I_3 are 0.564 and 0.087, respectively. We can now use (5.11) to give

---

(continued)

$$<i_n{}^2>_c = \left(S_I + \frac{S_E}{R_{in}{}^2}\right)BI_2 + (2\pi C_T)^2 S_E B^3 I_3$$

$$=2.82 \times 10^{-16} + 6.17 \times 10^{-17}$$

$$=3.44 \times 10^{-16} A^2.$$

These figures are typical for a receiver operating under these conditions. We can see that the series noise generator has a negligible effect on the receiver noise. However, we should note that, because of the B^3 dependency, this noise source becomes dominant at high data rates. A tenfold increase in B gives noise of 6.45×10^{-14} A^2.

To find the required optical power, we need to find the responsivity of the detector. From (4.13) we find that

$$R_0 = 0.62 \text{ A/W}$$

and so, from (5.27), we find that we need 180 nW (–37.46 dBm) of optical power for an error rate of 1 bit in 10^9 ($Q = 6$).

5.2.4   Photodiode Noise

As discussed in Chap. 4, the photodiode noise falls into two main categories – invariant dark current noise and signal-dependent shot noise. In a PIN receiver, the signal-dependent noise is often insignificant compared with the circuit noise, but with an APD receiver, we cannot ignore the signal noise. As we saw in Chap. 4, it is common practice to approximate the APD signal current to a Gaussian random variable. Hence the sensitivity analysis we have just done will be valid for an APD receiver.

In order to simplify the following work, we will assume that the extinction ratio is zero, that is, I_{min} is zero. So, by making use of the photodiode noise equations derived in Chap. 4, the noise current spectral density of an APD can be written as

$$S_{Id} = 2qI_{DB}M^2F(M) + 2qI_{DS} \qquad (5.28)$$

and

$$S_{IS} = 2q <I_s> M^2F(M) \qquad (5.29)$$

where $<I_s>$ is the average signal current. We can treat these noise sources in the same manner as the preamplifier shunt noise source. Thus, the equivalent input noise current due to the photodiode is

$$< i_n{}^2>_{\mathrm{pd}} = (S_{\mathrm{Id}} + S_{\mathrm{Is}})BI_2 \tag{5.30}$$

Now, $<I_s>$ is dependent on the presence of an optical pulse in a particular time slot. When a pulse is present in the time slot, $<I_s>$ is $I_{\max}$ while $<I_s>$ is zero when no pulse is present (assuming complete extinction of the source). So, the equivalent input noise currents for logic 1 and logic 0 pulses are

$$< i_n{}^2>_1 = 2qI_{\max}M^2F(M)BI_2 + < i_n{}^2>_{\mathrm{T}} \tag{5.31}$$

and

$$< i_n{}^2>_0 = < i_n{}^2>_{\mathrm{T}} \tag{5.32}$$

where $<i_n{}^2>_{\mathrm{T}}$ is the *total*, signal-independent, equivalent input noise current which includes the noise from the photodiode dark currents and any preamplifier noise. As the noise for logic 1 and logic 0 signals is different, the probability density plots of Fig. 5.7 will be different. If we account for this, and $I_{\min}$ is zero, Q will be given by

$$Q = \frac{I_{\max}}{\sqrt{< i_n{}^2>_1} + \sqrt{< i_n{}^2>_0}} \tag{5.33}$$

Thus the mean optical power required is

$$P = \frac{Q}{2MR_{\mathrm{o}}}\left(\sqrt{< i_n{}^2>_1} + \sqrt{< i_n{}^2>_0}\right) \tag{5.34}$$

If we substitute for $<i_n{}^2>_1$ and $<i_n{}^2>_0$, we get

$$\begin{aligned}
P &= \frac{Q}{2MR_{\mathrm{o}}}\left(\sqrt{2qI_{\max}M^2F(M)BI_2 + < i_n{}^2>_{\mathrm{T}}} + \sqrt{< i_n{}^2>_{\mathrm{T}}}\right) \\
&= \frac{Q}{2MR_{\mathrm{o}}}\left(\sqrt{4qPR_{\mathrm{o}}M^2F(M)BI_2 + < i_n{}^2>_{\mathrm{T}}} + \sqrt{< i_n{}^2>_{\mathrm{T}}}\right)
\end{aligned} \tag{5.35}$$

After some lengthy rearranging, we can express the sensitivity as

$$P = \frac{Q}{R_{\mathrm{o}}}\left(\frac{\sqrt{< i_n{}^2>_{\mathrm{T}}}}{M} + qBI_2QF(M)\right) \tag{5.36}$$

The first term in the brackets is inversely proportional to the avalanche gain, while the second term is, indirectly, dependent on M. Thus there must be an optimum value of M which minimises the required optical power. In order to find this optimum, we differentiate (5.36) with respect to M, equate the result to zero, and solve to find M_{opt}. (To simplify the derivation, we ignore the multiplied dark current. The effect of this approximation is not very dramatic.) Omitting the straightforward but lengthy mathematics, M_{opt} is given by

$$M_{opt} = \frac{1}{\sqrt{k}} \sqrt{\frac{\sqrt{<i_n^2>_T}}{qBI_2Q} + k - 1} \tag{5.37}$$

where k is the APD carrier ionisation ratio. In practice, because of all the approximations, (5.37) will only give an indication of the optimum gain. As we can vary the APD gain by altering the bias voltage, the optimum gain is often determined experimentally when the optical link is installed.

Example
In the optical receiver described in the previous example, the PIN detector is replaced by a silicon APD with the same quantum efficiency, a gain of 100, $k = 0.02$, and $I_{DB} = I_{DS} = 10$ nA. Determine the receiver sensitivity, the optimum APD gain and the sensitivity at this gain.
In order to find the receiver sensitivity, we must use (5.36):

$$P = \frac{Q}{R_o} \left(\frac{\sqrt{<i_n^2>_T}}{M} + qBI_2QF(M) \right)$$

For an error rate of 1 in 10^9 pulses, $Q = 6$. The responsivity of the APD is the same as the PIN detector used previously, that is, $R_0 = 0.62$ A/W. We also need to find the excess noise factor of the APD. By using (4.33) we find that $F(M) = 4$ for electrons producing avalanche gain. We also need to find the total signal-independent noise $<i_n^2>_T$.

If we assume that the source is totally extinguished, we can use (5.32) to give

$$<i_n^2>_T = <i_n^2>_0$$
$$= (2qI_{DB}M^2F(M) + 2qI_{DS})BI_2 + <i_n^2>_c$$
$$= 5.16 \times 10^{-16} + 2.02 \times 10^{-17}$$
$$= 5.36 \times 10^{-16} \text{ A}^2$$

From these figures we can see that the APD noise dominates over the preamplifier noise. This is due to the large amount of multiplied bulk leakage current. So, the required optical power is

(continued)

$$P = \frac{6}{0.62} \times \left(2.3 \times 10^{-10} + 0.22 \times 10^{-10}\right)$$
$$= 2.4 \text{ nW or} - 56.13 \text{ dBm}$$

To find the optimum avalanche gain, we use (5.37) to give

$$M_{\text{opt}} = \frac{1}{\sqrt{k}} \sqrt{\frac{\sqrt{<i_n^2>_T}}{qBI_2Q} + k - 1}$$
$$= \frac{1}{\sqrt{0.02}} \sqrt{\left(1.2 \times 10^3 - 0.98\right)}$$
$$= 244$$

The sensitivity of the receiver with this avalanche gain is

$$P = \frac{6}{0.62} \left(0.95 \times 10^{-10} + 0.22 \times 10^{-10}\right)$$
$$= 1.1 \text{ nW or} - 59.46 \text{ dBm}$$

From these calculations we can see that the use of an APD increases the sensitivity of the receiver. We can also see that, because of the small light levels, the signal-dependent APD noise is minimal when compared with the noise from the multiplied dark current. (In Sect. 5.4 we will compare PIN and APD receivers in greater detail.)

5.2.5 Timing Extraction

The sensitivity analysis just presented assumed central decision detection. As we saw earlier, the D-type flip-flop requires a clock of period equal to the time slot width. We could transmit this clock as a separate signal, but it is more usual to extract the clock from the received data. This is the function of the timing extraction circuit shown in Fig. 5.10.

The input to this circuit, taken from the threshold crossing detector, is first differentiated and then full-wave rectified. These two operations result in a series of pulses with the same period as that of the required clock signal. It is then a simple matter to use a phase-lock-loop, PLL, or a high Q tuned circuit, to extract the clock required by the decision gate.

So long as there are a large number of data transitions, the circuit will maintain a clock to the flip-flop. However, with the NRZ signalling format we are considering, a long sequence of 1s or 0s will cause a loss of the clock. This is because the PLL, or tuned circuit, will not receive any pulses. One solution to this problem is to use an alternative signalling format, such as *bi-phase* (or *Manchester*) coding. With this

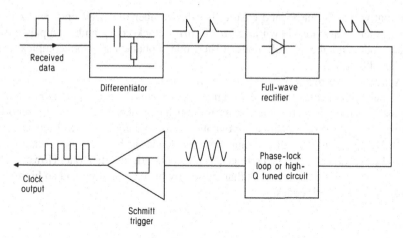

Fig. 5.10 Schematic diagram of a timing extraction circuit

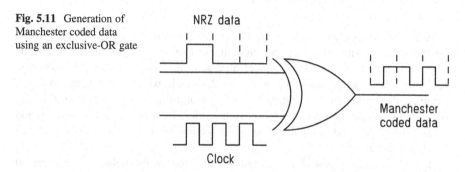

Fig. 5.11 Generation of Manchester coded data using an exclusive-OR gate

code, each time slot contains a data transition regardless of the logic symbol (Fig. 5.11), and this increases the timing content. The major disadvantage of this code is that the pulse width is half that of full-width pulses, resulting in a doubling of the required bandwidth.

Although bi-phase coding is often used in low bit-rate links and local area networks, *LANs,* the doubling in data-rate makes this format unattractive for use in high-speed telecommunications links. For this application, *block coding* of the NRZ data is used. With this type of coding, a look-up table converts m bits of input data into n bits of output data $(n > m)$. Such codes are known as *mBnB* block codes, and they enable designers to increase the timing content of NRZ signals by limiting the maximum number of consecutive 1s or 0 s.

Block coding of random data also helps to alleviate *base-line* wander, which causes the amplitude of a long sequence of like symbols to sag; in extreme circumstances, the amplitude of a long sequence of ones drops below the threshold level, and the error rate increases. Base-line wander is due to a poor receiver low-frequency response filtering out the strong d.c. content of a long sequence of ones or zeros. For block-coded data to exhibit zero d.c. content, we must limit the maximum number of consecutive like symbols and ensure that the number of coded ones and zeros is equal. Such codes are known as *zero disparity* block codes and examples are the 7B8B code and the 8B10B code. (The code alphabets are too complex to be presented here.)

A useful feature of line coded data is that the spectrum has a lower cut-off frequency below which there are no signal components. Hence *supervisory channels* can use the empty low-frequency spectrum. Such channels are required for reporting on the state of various system components and to send control data to repeaters and terminal equipment. One disadvantage of block codes is that, because of their inbuilt redundancy, the encoded data rate is n/m times the original data rate. However, this increase is significantly less than that caused by bi-phase coding and so high-speed links often use block codes.

5.3   Analogue Receiver Noise

Thus far we have only considered digital optical transmission links. However, some optical links transmit analogue information, for example, composite video signals and analogue information from optical fibre sensors. Consequently, this section is concerned with analogue receiver noise. Although we use the term analogue receiver, the only difference between analogue and digital receivers is in the way the signals are processed after the post-amplifier. Depending on the modulation format, there may be some form of pre-detection filter prior to recovery of the baseband signal.

Let us consider sinusoidal amplitude modulation of the light, with a received optical power, $p(t)$, given by

$$p(t) = P_r(1 + ms(t)) \tag{5.38}$$

where P_r is the average received optical power, $s(t)$ is the modulating signal and m is the modulation depth. For an APD, this signal produces a photodiode current, $i_s(t)$, given by

$$i_s(t) = R_0 M p(t) \tag{5.39}$$

and so the m.s. signal current, ignoring a constant d.c. term, is

$$< I_s^2 > = \frac{1}{2} (R_0 M m P_r)^2 \tag{5.40}$$

From our discussions about digital receiver noise, we can write the equivalent input m.s. noise current as

$$< i_n{}^2>_T = \int_0^{B_{eq}} 2qR_oP_rM^2F(M)df + \int_0^{B_{eq}} 2qI_{DB}M^2F(M)df + < i_n{}^2>_c \quad (5.41)$$

where B_{eq} is the noise equivalent bandwidth of the receiver, given by

$$B_{eq} = \int_0^\infty [H_T(\omega)]^2 df \quad (5.42)$$

Performing the integrations in (5.43) yields

$$< i_n{}^2>_T = 2q(I_{DB} + R_oP_r)M^2F(M)B_{eq} + < i_n{}^2>_c \quad (5.43)$$

Now, the preamplifier noise current is given by

$$< i_n{}^2>_c = \int_0^{B_{eq}} \left(S_I + \frac{S_E}{R_{in}{}^2} \right) df + \int_0^{B_{eq}} S_E(\omega C_T)^2 df$$

or

$$< i_n{}^2>_c = \left(S_I + \frac{S_E}{R_{in}{}^2} \right) B_{eq} + (2\pi C_T)^2 S_E \frac{B_{eq}{}^3}{3} \quad (5.44)$$

and so the signal-to-noise ratio is

$$\frac{S}{N} = \frac{< I_s{}^2 >}{< i_n{}^2>_T} = \frac{1}{2} \frac{(R_oMmP_r)^2}{2q(I_{DB} + R_oP_r)M^2F(M)B_{eq} + < i_n{}^2>_c} \quad (5.45)$$

As with the digital receiver, there is an optimum value of avalanche gain. We can find this value by differentiating (5.45) with respect to M, equating the result to zero and solving for M. Thus we can find the optimum gain from

$$M_{opt}{}^{2+x} = \frac{< i_n{}^2>_T}{q(I_{DB} + R_oP_r)B_{eq}} \quad (5.46)$$

where we have made use of $F(M) = M^x$ and $<i_n{}^2>_T$ is as defined by (5.43). As with a digital receiver, the theoretical value of M_{opt} is an indication of the optimum. When the receiver is commissioned, we can alter the APD bias to give the optimum S/N.

We have now completed our theoretical study of digital and analogue receivers. Before we go on to consider various preamplifier designs, we shall perform some sensitivity calculations.

5.4 Comparison of APD and PIN Receivers

In this section we will calculate the analogue and digital sensitivity of a receiver employing a PIN photodiode and compare it with that of the same receiver using n APD. We will assume the receivers to have a bandwidth of 100 MHz, which allows for the detection of 200 Mbit/s digital data with a NRZ format. We will consider two levels of preamplifier noise: 10^{-14} A^2, a somewhat noisy design, and 10^{-16} A^2, a typical state-of-the-art design.

The responsivity of both detectors will be taken as 0.5 A/W at 850 nm. We will assume that the noise from the PIN leakage current is negligible in comparison with other noise sources, while the APD surface and bulk leakage currents will be taken to be identical and equal to 10 nA. For the APD, we take a multiplication factor of 100 and an excess noise factor of 4. In order to simplify the work, we will assume that the source is completely extinguished.

If we initially consider the noisy preamplifier with a PIN photodiode, then use of (5.27) results in a sensitivity, for a 10^{-9} error rate, of

$$P = \frac{6}{0.5} \sqrt{10^{-14}} \, \text{W} = 1.2 \, \mu\text{W} = -29.21 \, \text{dBm}$$

If we replace the PIN detector by the APD, we must first calculate the total signal-independent shot noise. Thus

$$\begin{aligned}
<i_n{}^2>_T &= <i_n{}^2>_c + 2qI_{DS}I_2B + 2qI_{DB} \, M^2F(M)I_2B \\
&= 1 \times 10^{-14} + 3.61 \times 10^{-19} + 1.4 \times 10^{-14} \\
&= 2.4 \times 10^{-14} \, \text{A}^2
\end{aligned}$$

As can be seen, the surface leakage current shot noise is insignificant when compared with the bulk leakage current shot noise. By substituting $<i_n{}^2>$ into (5.38) we get

$$P = \frac{6}{0.5} \left(\frac{\left(2.4 \times 10^{-14}\right)^{1/2}}{100} + 4.3 \times 10^{-10} \right)$$
$$= 21 \, \text{nW or} - 46.8 \, \text{dBm}$$

Examination of Table 5.1 shows that the use of an APD with a noisy digital receiver results in a significant increase in sensitivity, compared with that obtained with a PIN detector. However, if the receiver noise is low, the advantage is considerably reduced. This is because the APD noise dominates the total receiver noise, and so any reduction in preamplifier noise will not produce a significant change in sensitivity. With a PIN detector, however, the preamplifier noise is

Table 5.1 Comparison of digital receiver performance using PIN and APD detectors. The terms in brackets are the avalanche gain of the APD. The second set of figures in the APD columns relate to the optimum avalanche gain

Detector	PIN		APD	
Preamplifier noise level (A^2)	10^{-14}	10^{-16}	10^{-14}	10^{-16}
Digital receiver sensitivity (dBm)	-29.21	-39.21	-46.80 (100)	-51.94 (100)
			-46.85 (215)	-52.00 (68)

dominant, and so a reduction in preamplifier noise causes a large change in sensitivity.

In the next section we will consider ways of measuring receiver sensitivity. Also presented is a method of determining the sensitivity from a knowledge of the output noise characteristic and the receiver transfer function.

5.5   Measurement and Prediction of Receiver Sensitivity

5.5.1   Measurement of Receiver Sensitivity

The sensitivity of an optical receiver (i.e., the preamplifier and associated signal processing circuitry) detecting digital data can be measured directly with an error-rate test set. This equipment comprises a pseudo-random binary sequence, PRBS, generator and an error-rate detector. The PRBS generator modulates a light source, the output of which is coupled to the receiver photodiode. The output of the receiver D-type flip-flop is then applied to the error detector. This instrument compares the detected PRBS with the transmitted sequence and counts the number of errors in a certain time interval. It is then a simple matter to find the probability of an error, P_e.

We can find the mean optical power resulting in the measured error rate by monitoring the photodiode current and then dividing by the responsivity. Attenuators placed in the optical path will vary the received power and hence the number of errors. If a graph of P_e against optical power is then plotted, the required power for a specified error rate can be easily found. (This graph will take the form of Fig. 5.8.)

As previously noted, the low-level light signal is unlikely to be zero, and so the ammeter monitoring the photodiode current will read $I_{max}/2$ and not $(I_{max} - I_{min})/2$. We can convert the ammeter reading into the current we require by using the following formula:

$$\frac{I_{max} - I_{min}}{2} = \frac{I_{max}}{2}(1 - \epsilon) \tag{5.47}$$

where ϵ is the extinction ratio.

The use of an error-rate test set allows us to measure the sensitivity of an optical receiver directly. An alternative method of predicting the sensitivity is based on measurements of the output noise and the transimpedance.

5.5.2 *Prediction of Receiver Sensitivity*

The theoretical prediction of receiver sensitivity relies on calculating the noise spectral density at the input to the preamplifier. We can make use of this to predict the sensitivity from a knowledge of the preamplifier output noise spectrum and the transimpedance.

Most modern spectrum analysers have a facility for measuring noise spectral density. We can find the output noise voltage spectral density of a preamplifier by boosting the output noise using a cascade of wideband amplifiers. If we divide this noise characteristic by the total transimpedance, we get the equivalent input noise current spectral density. It is then a simple matter to perform a curve fitting routine to determine the values of the frequency invariant and the f^2 variant noise current spectral density components. We can use these values in place of the analytical coefficients in (5.11) to determine the input noise current and hence the receiver sensitivity.

As an example, let us consider the output noise voltage spectral density shown in Fig. 5.12a. When divided by the transimpedance, Fig. 5.12b, the equivalent input noise current spectral density takes the form of Fig. 5.12c. From this figure, the frequency invariant coefficient is 4×10^{-24} A^2/Hz, while the f^2 variant term is approximately 2.1×10^{-40} A^2/Hz^3. Hence we can predict the sensitivity of this receiver from

$$P = \frac{Q}{R_0} \sqrt{4 \times 10^{-24} I_2 B + 2.1 \times 10^{-40} I_3 B^3} \qquad (5.48)$$

This method relies on an accurate knowledge of the receiver transimpedance. The low-frequency transimpedance can be obtained by injecting a constant current into the preamplifier. We can obtain this by connecting a sweep generator to the constant current source and monitoring the output on a spectrum analyser. An alternative is to modulate a light source with the output of the sweep generator and connect a spectrum analyser to the preamplifier output as before.

Apart from experimental errors, the major source of error results from the use of a theoretically ideal pre-detection filter. In spite of this, the predicted sensitivity can be within 1 dB of the actual receiver sensitivity.

Problems

1. Estimate the bandwidth of the optical receiver whose characteristics are shown in Fig. 5.12.

[200 MHz]

2. Predict the sensitivity of the receiver given in Fig. 5.12 at a data rate of 100 Mbit/ s. Take a responsivity of 0.5 A/W and a Q of 6 (corresponding to 1 error in 10^9 bits).

[37.3 dBm]

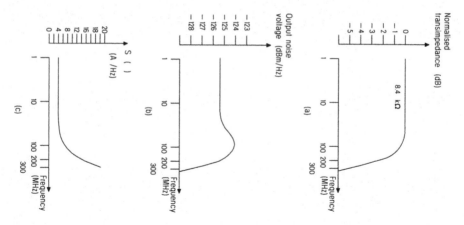

Fig. 5.12 (a) Transimpedance relative to mid-band; (b) output noise spectral density; (c) equivalent input noise current spectral density of a PINBJT transimpedance preamplifier. All graphs are based on experimental data

3. Show that the noise equivalent bandwidth of a single pole receiver with bandwidth f_o is given by $f_o\pi/2$.
4. Determine the analogue S/N ratio for the receiver of Fig. 5.12 using a PIN photodiode with responsivity 0.5 A/W. Take a bandwidth of 100 MHz; a received power of 1 µW; and a modulation depth of 0.3.

                                                                                              [12 dB]
5. Repeat Q4 if an APD is used with the same responsivity; a gain of 100; bulk leakage current of 10 nA; an excess noise factor of 4.

                                                                                              [52 dB]

Recommended Reading

1. Personick SD (1973) Receiver design for digital fiber optic communication systems, Parts I and II. Bell System Tech J 52:843–886
2. Smith DR, Garrett I (1978) A simplified approach to digital receiver design. Opt Quant Electron 10:211–221

Chapter 6
Preamplifier Design

In the previous chapter, we assumed that the detector-preamplifier combination, which we shall now call the receiver, had a bandwidth of at least 0.5 times the bit rate, or the baseband bandwidth for analogue signals, and low noise. In this chapter we will consider the design and analysis of various preamplifier circuits, with the aim of optimising these characteristics. We shall consider the two most common types of preamplifier – the *high input impedance* design and the *transimpedance* design. In the noise analyses presented, we will only consider the performance of preamplifiers receiving digital signals.

The high input impedance preamplifier is the most sensitive design currently available and, as such, finds applications in long-wavelength, long-haul routes. The high sensitivity is due to the use of a high input resistance preamplifier (typically >1 MΩ) which results in exceptionally low thermal noise. The high resistance, in combination with the receiver input capacitance, results in a very low bandwidth, typically <30 kHz, and this causes integration of the received signal; indeed, these receivers are commonly called *integrating front-end* designs. A differentiating, *equalising* or *compensating*, network at the receiver output corrects for this integration.

In contrast, the transimpedance design relies on negative feedback to increase the bandwidth of the open-loop preamplifier, and so a compensation circuit is not normally required. Although the resulting receiver is often not as sensitive as the integrating front-end design, this type of preamplifier does exhibit a high dynamic range.

Both types of preamplifier can use either field effect transistors, FETs, or bipolar junction transistors, BJTs, as the input device. FET input receivers are usually more sensitive than BJT input receivers; however, as we shall see later, the situation can change at data rates greater than 1 Gbit/s. When we examine the integrating front-end receiver, we will consider a FET input design, whereas when we consider a transimpedance receiver, we will use a BJT.

© Springer Nature Switzerland AG 2020
M. Sibley, *Optical Communications*, https://doi.org/10.1007/978-3-030-34359-0_6

6.1 High Input Impedance Preamplifiers

As these designs rely on a very high input resistance to produce a sensitive receiver, the choice of front-end transistor is important. BJTs have a relatively low input resistance and so are seldom used. On the other hand, FETs exhibit a very large input resistance, and so these are the obvious choice for the front-end device. Integrating front-end preamplifiers usually consist of a PIN photodiode feeding a FET input preamplifier. The resulting circuit is commonly known as a *PINFET* receiver, and a typical design is shown in simplified form in Fig. 6.1. (For reasons of clarity, we have not included the biasing components.)

The front-end is a common-source, *CS*, stage feeding a common-base, *CB*, stage (as shown in Fig. 6.1). This configuration, known as a *cascode* amplifier, results in a low input capacitance and a high voltage gain. (It is not necessary to use BJTs at all; some designs use FETs throughout and can be fabricated as gallium arsenide integrated circuits.)

Also shown in Fig. 6.1 is the compensation network which has a zero at the same frequency as the front-end pole and a pole which determines the receiver bandwidth. The 50 Ω load resistor across the output of the compensation network represents the input resistance of any following amplifier.

We shall now examine the frequency response (taking account of the compensating network), noise characteristic and dynamic range of a PINFET receiver. Although we only consider FET input designs, a similar analysis would also apply to PINBJT receivers.

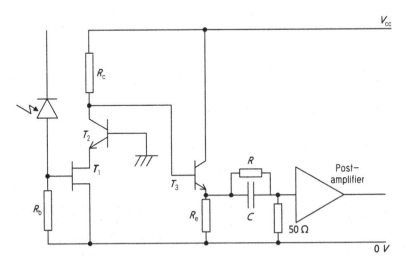

Fig. 6.1 A basic PINFET optical receiver with equalisation network and post-amplifier

6.1.1   Frequency Response

The time constants associated with the front-end and the cascode load determine the frequency response of the PINFET receiver shown in Fig. 6.1. The first time constant, τ_{in}, causes a pole which is usually located at about 30 kHz, and it is this that the compensating network must counteract. Such a low frequency is usually below the lower cut-off frequency of the data (line coded to reduce low frequency content).

We can determine the relevant time constants by drawing the a.c. equivalent circuit and then finding the resistance in parallel with each capacitance. Thus, with reference to Fig. 6.2, the front-end pole, s_1, is

$$s_1 = \frac{1}{\tau_{in}} \qquad (6.1)$$

where

$$\tau_{in} = R_b C_{in} \qquad (6.2)$$

Here C_{in} is the total input capacitance, which is the sum of the diode capacitance, C_d; the FET gate-source capacitance, C_{gs}; the stray input capacitance, C_s; and the *Miller* capacitance, $(1 - A_1)C_{gd}$. The parameter C_{gd} is the FET gate-drain capacitance, and A_1 is the voltage gain of the FET stage, given by

$$A_1 = -g_{m1} r_{e2}$$

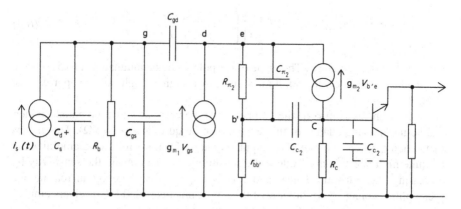

Fig. 6.2 A.c. equivalent circuit of a PINFET receiver

or

$$A_1 = -g_{m1} \frac{V_T}{I_{e2}} \tag{6.3}$$

where $V_T = 25$ mV and g_{m1} is the FET *transconductance*. Thus C_{in} will be given by

$$C_{in} = C_d + C_{gs} + C_s + \frac{(1 + g_{m1} V_T) C_{gd}}{I_{e2}} \tag{6.4}$$

Although the FET is usually biased at about 15 mA, g_{m1} is typically 15 mS (considerably lower than that achievable by a BJT operating at the same current). This is because, unlike a BJT, the g_m of a FET is relatively independent of bias current. This low g_m, together with the load resistance of r_{e2}, means that A_1 will be very low. However, the gain of the common-base stage, A_2, is $g_{m2} R_c$ where g_{m2} is the transconductance of the CB stage, and so the total voltage gain, A_0, may well be high ($A_0 = A_1 A_2$).

It is important to minimize C_{in} because this will allow for the use of a larger value of R_b, for the same pole location, and hence a reduction in thermal noise. As we shall see in the next section, a small input capacitance will also reduce the preamplifier noise.

To equalise for the front-end integration, the compensating network should have a zero at the same frequency as the input pole. We can find the location of the compensating zero by noting that the transfer function, $H_{eq}(\omega)$, of the compensation network is

$$H_{eq}(\omega) = \left[\frac{50}{50 + R} \right] \left[\frac{1 + j\omega CR}{1 + j\omega CR 50 / (50 + R)} \right] \tag{6.5}$$

and so CR should equal τ_{in}. The value of the pole in the denominator of (6.5) sets the bandwidth of the receiver, and, if the value of R is high enough, we can approximate it to $1/C50$.

At frequencies below s_1, the compensation network acts as a potential divider, and so the transimpedance of the receiver can be quite low (<10 kΩ). This means that, when referred to the receiver input, any noise from the following amplifier can be quite high and this may reduce the sensitivity. We can regain the sensitivity by increasing the transimpedance, and this is most easily done by increasing the preamplifier voltage gain. In view of this, most practical PINFET receivers employ additional, low-noise amplification stages prior to compensation.

If we compensate for the front-end pole, the next major pole, s_2, is that associated with the cascode load time constant, τ_c. If s_2 is lower in frequency than the compensation network pole, then the receiver will fail; thus it is important to determine t_c and hence s_2. By referring to Fig. 6.2, we can see that t_c approximates to

$$\tau_c = R_c 2C_c \quad \text{and so} \tag{6.6}$$

$$s_2 = \frac{1}{2R_c C_c} \tag{6.7}$$

where C_c is the BJT collector-base capacitance. This pole location is an approximation because we are neglecting the base-spreading resistance, $r_{bb'}$, and the loading effect of the output emitter follower.

6.1.2 Noise Analysis

In this preamplifier there are three sources of noise: thermal noise from R_b, thermal noise due to the channel conductance and shot noise from the gate leakage current, I_g. It may be recalled from the previous chapter that we need to find the total equivalent input noise current in order to determine the receiver sensitivity.

We can see from Fig. 6.3 that the I_g generator and the R_b generator are both connected to the input node and so are easily dealt with. However, we must refer the channel conductance thermal noise to the input by some means. To do this, we note that the $<i_n^2>_d$ generator produces a noise current spectral density of $4kT\Gamma g_m$ A^2/Hz, in a short circuit placed across the drain and source. (The parameter Γ is known as the FET constant, and it has an approximate value of 0.7 for Si and 1.1 for GaAs FETs.) We can refer this current to the input of the FET by dividing by the transconductance. So, a gate-source noise voltage generator of spectral density $4kT\Gamma/g_m$ V^2/Hz will produce a m.s. short-circuit output current equal to the channel

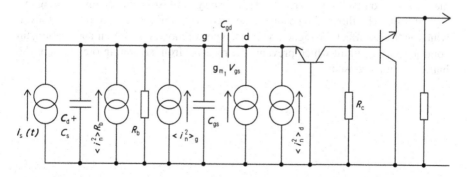

Fig. 6.3 Noise equivalent circuit of a PINFET receiver

conductance noise current. As this generator drives the input admittance of the short-circuit transistor, we can write the *total* equivalent input m.s. noise current, $<i_n^2>_c$, as

$$< i_n^2 >_c = \frac{4kTI_2B}{R_b} + 2qI_gI_2B + \frac{4kT\Gamma}{g_{m1}} \left\{ \frac{I_2B}{R_b^2} + (2\pi C_T)^2 I_3 B^3 \right\} \qquad (6.8)$$

where C_T is given by

$$C_T = C_d + C_{gs} + C_{gd} + C_s \qquad (6.9)$$

We should note that, as a result of short-circuiting the drain and source, the Miller capacitance does not appear in (6.9). We can simplify (6.8) if we assume that R_b is very large and the gate leakage current is very low. Thus, the receiver noise becomes

$$< i_n^2 >_{min} = \frac{4kT\Gamma}{g_{m1}} (2\pi C_T)^2 I_3 B^3 \qquad (6.10)$$

which represents the minimum amount of noise available from an ideal PINFET receiver. This clearly shows the need to minimise the input capacitance and use high g_m FETs.

6.1.3 Dynamic Range

The integration of the received signal at the front-end restricts the dynamic range of PINFET receivers – a long sequence of 1s will cause the gate voltage, V_g, to ramp up, and this may disrupt the biasing levels so causing the receiver to fail. In digital receivers, line-coded data can correct for this integration. For example, in the 5B6B code, the maximum number of consecutive ones will be six, and this will cause the gate voltage, V_g, to rise to a certain level. However, six zeros will eventually follow the six ones, to maintain a zero symbol disparity, and so V_g will ramp down again.

Unfortunately, line coding cannot take account of variations in input power level, which will also affect analogue receivers. The solution is to use an automatic gain control, or *agc*, circuit which prevents the receiver from saturating (i.e. it keeps the bias conditions constant).

6.1.4 Design Example

A PINFET receiver is to operate at a data rate of 1 Gbit/s. The input stage of the design is a cascode arrangement, with a bias current of 15 mA. The g_m of the GaAs FET at this bias current is 15 mS, and the gate leakage current is 15 nA. The receiver is to have a total voltage gain of 100. Microwave bipolar transistors, with a collector-base capacitance of 0.2 pF, are used after the FET input. The design is fabricated on a hybrid thick-film circuit, resulting in a total input capacitance of 0.5 pF. Complete the design of the receiver and estimate the receiver sensitivity assuming an ideal pre-detection filter and an error rate of 1 in 10^9 bits. Take a responsivity of unity.

As this is a PINFET design, the bias resistor on the gate of the FET front-end needs to be as high as possible. The highest practical resistance is 10 MΩ which results in a front-end pole at 32 kHz. The equalising network must compensate for this pole. From Eq. (6.5) we find

$$\frac{1}{2\pi RC} = 32 \times 10^3$$

and so

$$RC = 5 \ \mu s$$

We require a minimum equalised bandwidth of 500 MHz, and so the pole in (6.5) must be at this frequency. Thus

$$\frac{50 + R}{2\pi CR50} = 500 \times 10^6$$

and so

$$R = 37 \ \Omega \ \text{and} \ C = 15 \ \text{pf}$$

The design of the equaliser assumes that the receiver has a very wide bandwidth when the front-end pole is equalised. Thus, we need to examine the equalised receiver bandwidth. This bandwidth will be set by the time constant associated with the cascode load and the input capacitance of the following stage. The common-source front-end sees the input resistance of the common-base stage as a load. Taking a cascode bias current of 15 mA, and a FET transconductance of 15 mS, we can use Eq. (6.3) to give

$$A_1 = g_{m1} \frac{V_T}{I_{e2}}$$
$$= 25 \times 10^{-3}$$

As the total voltage gain should be 100, the CB stage needs a gain of 4×10^3 resulting in a load resistor of 6.7 kΩ. Such a high value of R_c may cause difficulties with biasing conditions and limit the compensated bandwidth. Thus we will take R_c equal to 400 Ω. This results in A_0 being 6, and so we must use further amplification stages with a combined gain of 17.

The microwave BJT transistors used on the design have a C_c of 0.3 pF. Thus we can use Eq. (6.6) to give a compensated bandwidth of

$$f_2 = \frac{1}{2\pi\tau_c}$$
$$= 995 \ \text{MHz}$$

This assumes that the stage following the cascode does not introduce any loading effects. As we saw earlier, a high voltage gain is important, and so a common-emitter amplifier could follow the cascode stage to increase the gain. However, this will tend to increase the capacitance seen by the cascode load, due to the Miller effect, and so limit the equalised bandwidth. Thus the cascode should be followed by an emitter follower, which will act as a buffer to feed further common-emitter gain stages.

We now need to examine the noise performance of this design. If we ignore the photodiode leakage current, then the total equivalent input m.s. noise current will be given by (6.8). Thus,

$$< i_n^2 >_c = \frac{4kTI_2B}{R_b} + 2qI_gI_2B + \frac{4kT\Gamma}{g_{m1}}\left\{\frac{I_2B}{R_b^2} + (2\pi C_T)^2 I_3 B^3\right\}$$
$$= 9.0 \times 10^{-19} + 2.7 \times 10^{-18} + 1.2 \times 10^{-18}(5.64 \times 10^{-6} + 856.7) = 1.0 \times 10^{-15} \ \text{A}^2$$

where we have taken C_T to be 0.5 pF and $\Gamma = 1.1$. This results in a sensitivity of -37.15 dBm for R_0 equal to 1 and an error rate of 1 in 10^9. As we can see from these figures, the noise from the bias resistor and the gate leakage current is not very significant when compared with the channel noise. This is because the channel noise becomes even more dominant as the bit rate increases, and so Eq. (6.10) will accurately predict $<i_n^2>_c$.

6.2 Transimpedance Preamplifiers

An ideal *transimpedance* amplifier supplies an *output voltage* which is directly proportional to the *input current* and independent of the source and load impedance. We can closely approximate the ideal amplifier by using negative feedback

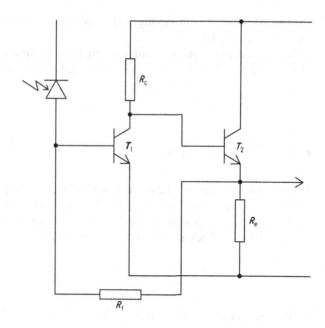

Fig. 6.4 A simple common-emitter/common-collector, shunt feedback transimpedance receiver

techniques to reduce the input impedance. If the open- loop amplifier is ideal, it has infinite input and zero output resistance, and so the transfer function of the feedback amplifier equals the impedance of the feedback network. As well as a predictable transfer function, a transimpedance preamplifier also exhibits a large closed-loop bandwidth, and so, in general, integration of the detected signal does not occur. Apart from the obvious advantage of not requiring a compensation network, the high bandwidth also results in a dynamic range which is usually larger than that of a PINFET receiver.

The choice of front-end transistor is entirely at the discretion of the designer; however, as we considered a FET input preamplifier previously, we will only examine BJT input transimpedance designs.

Figure 6.4 shows the circuit diagram of a simple, common-emitter (CE), common-collector (CC), shunt feedback preamplifier. Comparison with the PINFET receiver reveals that a feedback resistor, R_f, replaces the bias resistor and, by virtue of Miller's theorem, this resistance appears at the input as $R_f/1 - A_o$. Thus even if R_f is high, in order to reduce thermal noise, the input resistance will be less than that of the PINFET, and this will result in a higher bandwidth. (A_o is the *open-loop* voltage gain which we can find by breaking the feedback loop, loading the circuit with R_f at both ends of the loop and then calculating the voltage gain. Fortunately, the open-loop voltage gain is almost the same as the closed-loop voltage gain, and we shall use this approximation.)

In this design, A_o is the product of the front-end gain and the second-stage attenuation (which we shall assume to be negligible). As we shall see later, A_o should be as high as possible to achieve a large bandwidth; however, a high value of voltage gain may cause instability, and so most transimpedance designs have voltage gains of less than 100.

We will now proceed to examine the frequency response and noise performance of this design. Although we will consider a CE/CC shunt feedback preamplifier, the same methods can be applied to other designs.

6.2.1 Frequency Response

We can find the transfer function of a transimpedance preamplifier by applying standard feedback analysis using *impedances* rather than voltages. Thus, we can relate the closed-loop transfer function, $Z_c(s)$, to the open-loop transfer function, $Z_o(s)$, and the feedback network transfer function, $Z_f(s)$, by

$$\frac{1}{Z_c(s)} = \frac{1}{Z_o(s)} - \frac{1}{Z_f(s)} \qquad (6.11)$$

The open-loop transimpedance is given by

$$Z_o(s) = A_o(s) \times \frac{R_{in} R_f}{R_{in} + R_f}$$

where $A_o(s)$ signifies that A_o is frequency dependent. In order to simplify the mathematics, we shall assume that the input resistance, R_{in}, is high. Therefore

$$Z_o(s) = A_o(s) R_f \qquad (6.12)$$

From our discussion of the PINFET cascode receiver, it should be apparent that $A_o(s)$ has two *open-loop* poles: one associated with the input time constant, τ_{in}, and one due to the time constant of the CE stage load, τ_c. If we assume that $\tau_{in} \gg t_c$ (i.e. the front-end pole is dominant), we can write $A_o(s)$ as

$$A_o(s) = \frac{A_o}{1 + s\tau_{in}} \qquad (6.13)$$

where A_o is $-g_{m1} R_c$ and

$$\tau_{in} = R_f(C_d + C_s + C_f + C_{\pi 1} + (1 - A_o)C_{c1}) \qquad (6.14)$$

(The inclusion of the parasitic feedback capacitance, C_f, in (6.14) arises from the use of the *open-loop* time constant; that is, the feedback network is placed across the input node.) Thus $Z_o(s)$ is

$$Z_o(s) = \frac{A_o R_f}{1 + s\tau_{in}} \tag{6.15}$$

Also, $Z_f(s)$ is given by

$$Z_f(s) = \frac{R_f}{1 + s\tau_f} \tag{6.16}$$

where τ_f is the feedback circuit time constant, $R_f C_f$. So, we can write (6.11) as

$$\frac{1}{Z_c(s)} = \frac{1 + s\tau_{in}}{A_o R_f} - \frac{1 + s\tau_f}{R_f} \tag{6.17}$$

or

$$Z_c(s) = \frac{A_o R_{eff}}{1 + sR_{eff}(C_d + C_s + C_{\pi 1} + (1 - A_o)(C_{c1} + C_f))} \tag{6.18}$$

$$\text{where} \quad R_{eff} = \frac{R_f}{1 + A_o} \tag{6.19}$$

It is interesting to note that if A_o is large enough, (6.18) will reduce to

$$Z_c(s) = \frac{R_f}{1 + sR_f(C_{c1} + C_f)} \tag{6.20}$$

which is the transimpedance for an ideal amplifier. In practice, this condition is difficult to achieve. This is because a large voltage gain may cause the preamplifier to become unstable, owing to the movement of the closed-loop poles within the feedback loop.

We could have obtained Eq. (6.18) directly by applying Miller's theorem to the feedback loop. However, if the receiver has two significant poles within the feedback loop, that is, τ_c is not $\ll\tau_{in}$, then we must perform the previous analysis with the two-pole version of A_o. So,

$$A_o(s) = \frac{A_1}{(1 + s\tau_{in})} \times \frac{A_2}{(1 + s\tau_c)} \tag{6.21}$$

We will return to this point when we consider common-collector input preamplifiers.

6.2.2 Noise Analysis

If a transimpedance preamplifier has a FET input stage, then the noise characteristic will be the same as for the PINFET, provided we replace R_b by R_f in (6.8). However, the transistor noise sources for a BJT are the base current shot noise, $2qI_b$ A^2/Hz; the base-spreading resistance thermal noise, $4kTr_{bb'}$ V^2/Hz; and the collector current shot noise, $2qI_c$ A^2/Hz.

The base current shot noise and the feedback resistor thermal noise appear as current generators connected to the input node and so can be easily accounted for. In addition, we can treat the collector current shot noise in a similar manner to the channel noise of the FET. However, the $r_{bb'}$ noise generator is a series generator, and so we must divide by the source impedance to convert it to a current generator. Thus, if we neglect r_π, we can write the total equivalent input m.s. noise current for a digital receiver as

$$<i_n^2>_c = \frac{4kTI_2B}{R_f} + 2qI_bI_2B + \frac{2qI_{c1}}{g_{m1}^2}\left\{\frac{I_2B}{R_f^2} + (2\pi C_T)^2I_3B^3\right\} + 4kTr_{bb'}\left\{\frac{I_2B}{R_f^2} + (2\pi C_1)^2I_3B^3\right\}$$

(6.22)

where $C_1 = C_d + C_s + C_f$. If R_f is made very large, so that we can neglect its noise, (6.22) becomes

$$<i_n^2>_c = 2qI_bI_2B + \frac{2qI_{c1}}{g_{m1}^2}(2\pi C_T)^2I_3B^3 + 4kTr_{bb'}(2\pi C_1)^2I_3B^3$$

(6.23)

which represents the minimum noise in a bipolar digital receiver, be it a transimpedance design or an integrating front-end design. Comparison with the corresponding PINFET Eq. (6.10) shows that there are two extra terms: the I_b noise term and the $r_{bb'}$ term. We can only reduce these terms by employing high gain, low $r_{bb'}$ transistors.

It is interesting to note that the I_b shot noise term is proportional to I_c, while the collector current shot noise term is inversely proportional to I_e. (This can best be seen by substituting for g_m as I_c/V_T. Thus, there should be an optimum value of collector current, $I_{c,opt}$, that minimises the total noise. We can find this optimum by differentiating (6.22) with respect to I_c and equating the result to zero. Hence, $I_{c,opt}$ is given by

$$I_{c,opt} = 2\pi V_T C_T \beta^{\frac{1}{2}} B(I_3/I_2)^{\frac{1}{2}}$$

(6.24)

For analogue receivers, the equivalent equation is

$$I_{c,opt} = 2\pi V_T C_T \beta^{\frac{1}{2}} B_{eq}/\sqrt{3} \tag{6.25}$$

(It should be noted that we are assuming C_T to be independent of bias. In reality, $C_{\pi 1}$ varies with bias, and so we have to find $I_{c,opt}$ by constant iteration.) If we substitute (6.24) back into (6.23), then the minimum noise from a bipolar front-end preamplifier will be

$$<i_n^2>_{c,min} = (8\pi kT)\left(C_T/\beta^{\frac{1}{2}}\right)(I_2 I_3)^{\frac{1}{2}}B^2 + 4kTr_{bb'}(2\pi C_1)^2 I_3 B^3 \tag{6.26}$$

Comparison with the minimum noise from a PINFET receiver, (6.10), shows that, *provided the $r_{bb'}$ noise is insignificant*, the m.s. noise from a BJT front-end receiver increases as the square of the data rate, whereas the m.s. noise from a PINFET receiver increases as the cube of the data rate. So, although a PINFET receiver may be more sensitive than a BJT receiver at relatively low data rates, at high data rates (typically >1 Gbit/s), the BJT receiver can be more sensitive. The f_T, β and $r_{bb'}$ of the transistor will determine the exact crossover point.

6.2.3   Dynamic Range

If the bandwidth of a transimpedance preamplifier is high enough so that no integration takes place, then the dynamic range can be set by the maximum voltage swing available at the preamplifier output. As the output stage is normally an emitter follower, running this stage at a high current will increase the voltage swing.

If the final stage is not the limiting factor, the dynamic range will be set by the maximum voltage swing available from the gain stage. With the design considered, the collector-base voltage of the front-end is a V_{be} (0.75 V) when there is no diode current. In a digital system, this results in a maximum peak voltage of approximately 1 V. The optical power at which this occurs can be easily found by noting that the peak signal voltage will be $I_{max}R_{eff}A_o$. Thus it is a simple matter to find the maximum input current and hence the maximum optical power. In most practical receivers, the dynamic range is greater than 25 dB.

6.2.4 Design Example

A bipolar transimpedance receiver is constructed using state-of-the-art surface mount components on a p.c.b. The diode capacitance is dominated by the package and has a value of 0.8 pF. The microwave transistors used in the design have a collector-base capacitance of 0.3 pF, an f_T of 4 GHz, a current gain of 120 and an $r_{bb'}$ value of 10 Ω. The front-end transistor is biased with 2 mA of collector current. Complete the design and estimate the sensitivity if the receiver is to detect 140 Mbit/s data with an error rate of 1 in 10^9 bits.

In order to find the receiver bandwidth, we must initially determine the location of the open-loop poles. Let us first take a front-end voltage gain of 20 and a feedback resistance of 4 kΩ. As the receiver is fabricated on a p.c.b. using surface mount components, the parasitic capacitance associated with the feedback resistor is 0.1 pF. With these parameters, we find that the front-end time constant is, from Eq. (6.14)

$$\tau_{in} = R_f(C_d + C_s + C_f + C_{\pi 1} + (1 - A_o)C_{c1})$$
$$= 48 \text{ ns}$$

The second pole is due to the collector load of the front-end transistor interacting with the input impedance of the following stage. We require a voltage gain of 20 from the front-end, and so the value of the collector resistor must be 250 Ω with a bias current of 2 mA. As the second stage is an emitter follower, the input resistance will be very high, while the input capacitance will be 0.3 pF. Thus, the second pole will have a time constant given by

$$\tau_c = 250 \times 0.3 \times 10^{-12} = 75 \text{ ps}$$

As this is far lower than the front-end time constant, we can assume that the design has a single-pole transfer function. (In reality we should use the two-pole form and then determine whether the response is single pole in form. We will assume that this has been done!) We can use Eq. (6.18) to give the bandwidth of the receiver as

$$f_{3dB} = \frac{1}{2\pi R_{eff}(C_d + C_s + C_{\pi 1} + (1 - A_o)(C_{c1} + C_f)}$$
$$= 70 \text{ MHz}$$

As the receiver is to detect 140 Mbit/s data, this bandwidth is correct. If we design the receiver to have a very high voltage gain, we can use Eq. (6.20) which gives a bandwidth of 400 MHz. This clearly shows the desirability of a high voltage gain.

However, we should remember that this preamplifier has a two-pole response, and so stability requirements may limit the maximum voltage gain.

Let us now examine the noise performance of this design. The equivalent input noise current is, from Eq. (6.22)

$$< i_n^2>_c = \frac{4kTI_2B}{R_f} + 2qI_bI_2B + \frac{2qI_{c1}}{g_{m1}^2}\left\{\frac{I_2B}{R_f^2} + (2\pi C_T)^2 I_3B^3\right\} + 4kTr_{bb'}\left\{\frac{I_2B}{R_f^2} + (2\pi C_1)^2 I_3B^3\right\}$$
$$= 3.3 \times 10^{-16} + 4.2 \times 10^{-16} + 6.7 \times 10^{-22} + 2.0 \times 10^{-18}$$
$$= 7.52 \times 10^{-16} A^2$$

This results in a sensitivity of −37.8 dBm. It is interesting to note that the collector current shot noise is insignificant in comparison with the base current shot noise. This is due to the front-end current being well above the optimum value. From (6.24) this optimum current is 0.4 mA and, if we use this value, $<i_n^2>_c$ is $4.8 \times 10^{-16} A^2$ – a sensitivity of −38.8 dBm. (The change is not very dramatic because the R_f noise is dominant.) As noted previously, we can only obtain the optimum collector current by repeated calculation.

By way of comparison, an optimally biased BJT in the integrating front-end receiver previously considered would produce a sensitivity of −41.10 dBm (about 9 dB less than the FET receiver). However, at a data rate of 1 Gbit/s, the difference in sensitivity reduces to 4.6 dB, which clearly shows that BJT receivers will have an advantage at high data rates.

6.3 Common-Collector Front-End Transimpedance Designs

The major disadvantage of CE input designs is that, by virtue of the gain between the collector and base of the front-end transistor, the collector-base capacitance appears as a large capacitance at the input node. A high input capacitance implies a low feedback resistance (to obtain a high band-width), and so the thermal noise will be high. However, we can use a cascode input, which has a very low input capacitance, to get a large bandwidth. Unfortunately, cascode designs require a voltage reference to bias the CB stage correctly, and this can lead to a more complicated design.

One way of eliminating the input Miller capacitance is to use a common-collector (CC) input. As CC stages have a very high input resistance, preamplifiers using this input configuration will be a better approximation to the ideal amplifier than those using CE input stages. Unfortunately, CC stages do not exhibit voltage gain and so, as Fig. 6.5 shows, an amplifying stage has to follow the front-end.

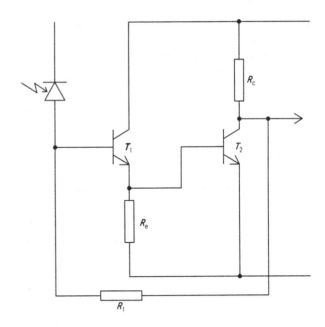

Fig. 6.5 A simple
common-collector/
common-emitter, shunt
feedback transimpedance
receiver

6.3.1   Frequency Response

The frequency response of CC input receivers is generally dominated by two poles:
one is due to the front-end, while the other is due to the input time constant of the CE
stage. By following a similar analysis to that used with the previous transimpedance
design, we can write $A_o(s)$ as

$$A_o(s) = \frac{A_1}{(1 + s\tau_{in})} \times \frac{A_2}{(1 + s\tau_c)} \tag{6.27}$$

where τ_{in} is

$$\tau_{in} = R_f(C_d + C_s + C_f + C_{c1}) \tag{6.28}$$

and τ_c is given by

$$\tau_c = R_{C\pi2}(C_{\pi2} + (1 - A_2)C_{c2}) \tag{6.29}$$

Here $R_{C\pi2}$ is the resistance in parallel with $C_{\pi2}$, given by

$$\frac{(R_{o1} + r_{bb'2})r_{\pi2}}{R_{o1} + r_{bb'2} + r_{\pi2}} \tag{6.30}$$

where R_{o1} is the open-loop, output resistance of the front-end. Thus $Z_c(s)$ is given by

$$Z_c(s) = \frac{-A_o R_{eff}}{1 + sR_{eff}(C_d + C_s + C_{c1} + (1 - A_o)C_f + \tau_c/R_f) + s^2 R_{eff} C_{in} \tau_c} \quad (6.31)$$

If the front-end pole is dominant, (6.31) simplifies to

$$Z_c(s) = \frac{-A_o R_{eff}}{1 + sR_{eff}(C_d + C_s + C_{c1} + (1 - A_o)C_f)} \quad (6.32)$$

Comparison with the equivalent equation for the CE design, (6.18), shows that the capacitive term is reduced. Therefore a CC input amplifier will have a greater R_f value than a CE design with the same bandwidth. Exactly the same conclusion applies to common-source FET input transimpedance preamplifiers.

We should note that $r_{bb'}$ affects the location of the second-stage pole – a high $r_{bb'}$ value results in a large $R_{C\pi2}$ and hence a large τ_c. A large τ_c results in s_2 being low in frequency, and this may cause the preamplifier transient response to exhibit undesirable over- and under-shoots. Hence a low $r_{bb'}$ will benefit both the receiver transfer function and noise.

6.3.2   Noise Analysis

The noise performance of a CC stage is similar to that of a CE stage. However, because the voltage gain of a CC stage is unity or less, there will be some noise from the second-stage $r_{bb'}$. This additional noise term, $<i_n^2>_2$, is given by

$$<i_n^2>_2 = \frac{4kT r_{bb'2}}{A_1^2} \left\{ \frac{I_2 B}{R_f^2} + (2\pi C_1)^2 I_3 B^3 \right\} \quad (6.33)$$

where A_1 is the voltage gain of the CC stage. Adding this to the terms in (6.22) gives the total equivalent input m.s. noise current. Again, we see the importance of using low $r_{bb'}$ transistors.

In conclusion, although there is an extra noise term with CC input preamplifiers, these designs do allow for the use of a greater value of R_f, and this may produce a net reduction in noise in comparison with a CE input design.

6.3.3   Design Example

The bipolar transimpedance receiver of the previous example is now designed to use a common-collector front-end. The same components and fabrication methods are employed as previous. Complete the design and estimate the receiver sensitivity.

In order to make a fair comparison with the CE design examined previously, we will use a CC input receiver with the same voltage gain and transistor parameters as before. So, for a voltage gain of 20, we require the collector load of the second stage to be 250 Ω for a bias current of 2 mA. Now, the resistance in parallel with $C_{\pi2}$ is, Eq. (6.30)

$$
\begin{aligned}
R_{C\pi2} &= \frac{(R_{o1} + r_{bb'2})r_{\pi2}}{R_{o1} + r_{bb'2} + r_{\pi2}} \\
&= \frac{(138 + 10)1.5 \times 10^3}{138 + 10 + 1.5 \times 10^3} \\
&= 135 \ \Omega
\end{aligned}
$$

Thus τ_c is, Eq. (6.29)

$$
\begin{aligned}
\tau_c &= R_{C\pi2}(C_{\pi2} + (1 - A_2)C_{c2}) \\
&= 135(2.9 + 6.3) \times 10^{-12} \\
&= 1.2 \text{ ns}
\end{aligned}
$$

We know that the receiver is likely to have a two-pole response. In order to determine the value of the feedback resistor, we must use Eq. (6.31) repeatedly until we get the required bandwidth. If this is done, we find that $R_f = 18$ kΩ for a 70 MHz bandwidth. So, as a check, Eq. (6.31) gives

$$
\begin{aligned}
Z_c(s) &= \frac{-A_o R_{eff}}{1 + sR_{eff}(C_d + C_s + C_{c1} + (1 - A_o)C_f + \tau_c/R_f) + s^2 R_{eff} C_{in} \tau_c} \\
&= \frac{-18 \times 10^3}{1 + s857(0.8 + 0 + 0.3 + 2.1 + 0.07) \times 10^{-12} + s^2 1.2 \times 10^{-18}} \\
&= \frac{-18 \times 10^3}{1 + s2.7 \times 10^{-9} + s^2 1.2 \times 10^{-18}}
\end{aligned}
$$

We can find the bandwidth by solving the quadratic in the denominator of this equation. Thus the receiver has two real poles at 74 and 284 MHz.

As regards the noise performance of this design, we use Eq. (6.22) together with Eq. (6.33) to give

Fig. 6.6 Bootstrapping of the photodiode capacitance in a CC-CE preamplifier

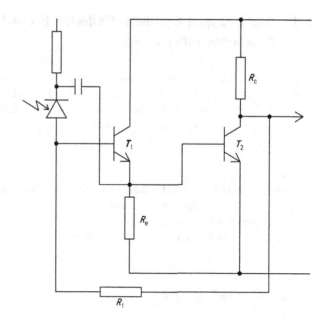

$$<i_n{}^2>_c = \frac{4kTI_2B}{R_f} + 2qI_bI_2B + \frac{2qI_{c1}}{g_{m1}{}^2}\left\{\frac{I_2B}{R_f{}^2} + (2\pi C_T)^2I_3B^3\right\}$$

$$+ 4kTr_{bb'}\left\{\frac{I_2B}{R_f{}^2} + (2\pi C_1)^2I_3B^3\right\} + \frac{4kTr_{bb'2}}{A_1{}^2}\left\{\frac{I_2B}{R_f{}^2} + (2\pi C_1)^2I_3B^3\right\}$$

$$= 0.7 \times 10^{-16} + 4.2 \times 10^{-16} + 6.7 \times 10^{-22} + 2.0 \times 10^{-18} + 2.2 \times 10^{-18}$$

$$= 4.9 \times 10^{-16}\,\text{A}^2$$

This results in a sensitivity of -38.75 dBm (an increase over the CE design of 0.77 dB). If we bias the front-end at the optimum collector current, then the increase in sensitivity is 1.7 dB, compared with an increase of only 1 dB for the CE input design. We can account for this difference by noting that the R_f noise is more dominant in the CE design and this tends to mask the advantage. A further advantage of CC input preamplifiers is that, unlike CE input designs, they generally maintain a flat frequency response when optimally biased.

6.4   Bootstrapped Common-Collector Front-End Transimpedance Designs

A useful modification to the common-collector front-end design is the implementation of bootstrapping the photodiode capacitance. The technique is shown in Fig. 6.6 in which the coupling capacitor is typically 0.1 μF and the bootstrapping resistor is 10 kΩ. The effect of this feedback is to change the photodiode capacitance as

$$C_{\mathrm{db-s}} = (1 - A_1)C_{\mathrm{d}} \tag{6.34}$$

The gain of a CC stage is theoretically one, but in practice it is typically 0.7 or 0.8. This results in a big reduction in C_{d} and hence an increase in bandwidth, significant if large area photodiodes are considered. (Free-space links use large area photodiodes. Such links are considered in the next chapter.)

Recommended Reading

1. Hooper RC, et al. (1980) PIN-FET hybrid optical receivers for longer wavelength optical communications systems, In: Proceedings of the 6th European Conference on optical communications, York, p 222–225
2. Smith DR et al (1980) PIN-FET hybrid optical receiver for 1.1-1.6 μm optical communication systems. Electron Lett 16:750–751
3. Hullett JL, Muoi TV, Moustakas S (1977) High-speed optical preamplifiers. Electron Lett 13:668–690
4. Sibley MJN, Unwin RT, Smith DR (1985) The design of PIN-bipolar transimpedance preamplifiers for optical receivers. J Inst Electr Electron Eng 55:104–110
5. Moustakas S, Hullett JL (1981) Noise modelling for broadband amplifier design, IEE Proceedings Part G: Electronic Circuits and Systems, vol 128, p 67–76
6. Millman J, Halkias C, Parikh CD (2017) Millman's integrated electronics, 2nd edn. Mc Graw-Hill, New York

Chapter 7
Current Systems and Future Trends

In previous chapters, we concentrated on the design and performance of individual components for use in optical links. What we have not yet examined is the overall design of practical links, and it is this that initially concerns us here.

When designing an optical link, system designers commonly use a *power budget* table which details the power losses encountered from source to receiver. This table enables the designer to implement system margins to account for ageing effects in the links. We will use the power budget to contrast two general cases: a low-speed data link using PCS fibre and LEDs and a high-speed telecommunications link using all-glass fibre and lasers. We will then examine the design of some current optical communications links.

In the rest of this chapter, we will examine some advanced components and systems that are being developed in the laboratory and assess their likely impact on optical communications.

7.1 System Design

The examples we consider here are an 850 nm wavelength, 10 Mbit/s link, operating over 500 m, and a long-haul, 1.55 μm wavelength, 10 Gbit/s link. As the length of the long-haul route has not been specified, we will use the power budget table to determine the repeater spacing. (Although these examples are tutorial in form, they will illustrate the basic principles behind link design.)

Table 7.1 shows the power budget for the two links. Because the short-haul link operates at a low data rate, we can specify PCS fibre and an LED source. We will also assume that the link is made up of five, 100 m lengths of fibre, requiring four pairs of connectors. For the high-speed link, we will take laser diode sources and 1 km lengths of dispersion-shifted, single-mode, all-glass fibre, connected together with fusion splices.

© Springer Nature Switzerland AG 2020
M. Sibley, *Optical Communications*, https://doi.org/10.1007/978-3-030-34359-0_7

Table 7.1 Link power budgets for a short-haul, low-data-rate link and a long-haul, high-data-rate link

	Short-haul link (10 Mbit/s)		Long-haul link (10 Gbit/s)	
Launch power	LED	−15 dBm	Laser	0 dBm
Receiver sensitivity		−45 dBm		−25 dBm
Allowable loss		30 dB		22 dB
Source coupling loss		3 dB		0.1 dB
Fibre loss	(6 dB/km)	3 dB		0.2 dB/km
Joint loss	(2 dB/pair)	8 dB		0.1 dB/splice
Detector coupling loss		3 dB		0.1 dB
Operating margin		13 dB		11 dB

The last parameter in the table, *operating margin*, represents excess power in the link. This excess is included to allow for the insertion of extra connectors, or splices, should a break in the fibre occur, as well as accounting for any power changes due to the effect of age. The operating margin may also be taken up by manufacturing variations in source power, receiver sensitivity and fibre loss.

We can find the distance between repeaters in the long-haul link by noting that the number of fibre-fibre splices is one less than the number of fibre sections. It is a simple matter to show that the maximum number of fibre lengths is 40, with 39 fusion splices. Thus the maximum length between repeaters is 40 km.

Although the power budget gives an indication of the maximum link length, it does not tell us whether the links can transmit the required data rate. From our previous discussions, the system bandwidth up to the input of the pre-detection filter, f_{3dB}, should be at least half the data rate, that is

$$f_{3dB} > \frac{B}{2} \tag{7.1}$$

We can relate this bandwidth to the rise time of the pulses, τ, at the input to the filter using

$$f_{3dB} = \frac{0.35}{\tau} \tag{7.2}$$

(Although this equation only applies to a network with a single-pole response, the error involved in the general use of (7.2) is minimal.) If we combine (7.1) and (7.2), the minimum rise time is given by

$$\tau \leq \frac{0.7}{B} \tag{7.3}$$

We can find the system rise time by adding the rise times of individual components on a mean square basis, that is

$$\tau^2 = \Sigma \tau^2_n \qquad (7.4)$$

(This equation results from convolving the impulse response of the individual components, to find the overall impulse response and hence the rise-time.) Most sources are characterised by the rise time of the optical pulses, while the receiver bandwidth is often quoted. However, as we saw in Chap. 2, optical fibre is often characterised by the pulse dispersion, and the impulse response can take on several different shapes. If we assume a Gaussian shape impulse response, then the rise time can be approximated by

$$\tau_{\text{fibre}} \approx 2.3\sigma \qquad (7.5)$$

where σ is the total fibre dispersion. So, if we return to the short-haul link, a fibre bandwidth of 35 MHz km (optical) gives a dispersion of 2.7 ns for a 500 m length, resulting in a rise time of 6.2 ns. (This assumes that the bandwidth is limited by modal dispersion. Hence, we can neglect the linewidth of the LED.) A 10 ns LED rise time and a receiver bandwidth of 10 MHz gives a total system rise time of 37 ns. From (7.3), this results in a maximum data rate of about 20 Mbit/s, and so the link is adequate for the 10 Mbit/s transmission speed that we require.

For the long-haul route, we will assume that the total fibre dispersion is 1.2 ps/nm/km. A laser linewidth of 0.1 nm yields a dispersion of 4.8 ps for the 40 km link length. This results in a rise time of approximately 11 ps which, together with a laser rise time of 15 ps and a receiver bandwidth of 5 GHz, yields a system rise time of 72.4 ps. The pulse width of 10 Gbit/s data is 100 ps, and so the link will transmit the required data rate.

These results, together with the link budget, indicate that even if we cut the operating margin on the long-haul route, the link could not be extended very far because of dispersion effects. Under these conditions, the link is said to be *dispersion limited*. However, the length of the short-haul route is determined by attenuation, that is, the link is *attenuation limited*, and so the link could be extended by reducing the operating margin. We should note that, because of the approximations involved in the calculation of link capacity, the actual data rates that can be carried are greater than those indicated. Hence the use of these formulae already allows an operating margin with both bit rate and attenuation.

7.2 Current Systems

In this section we will briefly examine the first optical transatlantic cable, *TAT8* (1988), and compare it to a state-of-the-art transoceanic link (2017). We will see that TAT8 uses technology that is still in use today, whereas modern long-haul links use advanced optical components that are, at present, too expensive or not necessary for short-haul links. We will then discuss a local area network, LAN, using PCS fibre.

TAT8 was the first optical transatlantic communications link. The cable was laid in three sections, with a different manufacturer taking responsibility for each. The first section was designed by the American Telephone and Telegraph Co., *AT&T*, and is a 5600 km length from the USA to a branching point on the continental shelf to the west of Europe. At this point, the cable splits to France and the UK. The French company *Submarcom* designed the 300 km length link to France, while the British company Standard Telephones and Cables, *STC*, were responsible for the 500 km link to the UK.

In view of the link length, dispersion effects are highly important, and so the operating wavelength is 1.3 μm. The use of single-mode laser diodes yields a total fibre dispersion of 2.8 ps/nm/km, and so the regenerator spacing is limited by fibre attenuation, not dispersion. The InGaAsP laser diode sources launch a minimum of approximately −6 dBm into single-mode fibre. With an average receiver sensitivity of −35 dBm for a 10^{-9} error rate, the allowable loss over a repeater length is 29 dB. A typical operating margin of 10 dB and a fibre attenuation of 0.48 dB/km result in a repeater spacing of 40–50 km.

At each repeater, PIN photodiodes feed BJT transimpedance preamplifiers. The signals are then amplified further, prior to passing through pre-detection filters to produce raised-cosine spectrum pulses. As the preamplifier is a transimpedance design, front-end saturation does not occur, and so an mBnB line code does not have to be used. Instead, the TAT8 system uses an even parity code, with a parity bit being inserted for every 24 transmission bits (a *24B1P* code). As this code has a low timing content, surface acoustic wave (*SAW*) filters with a *Q* of 800 are used in the clock extraction circuit. Should the timing circuit fail in a particular regenerator, provision is made for the data to be sent straight through to the output laser, so that the data can be retimed by the next repeater.

The TAT8 cable comprises six individual fibres; two active pairs carry two-way traffic at a data rate of 295.6 Mbit/s on each fibre. As well as parity bits, some of the transmitted bits are used for system management purposes, and so the total capacity is 7560 voice channels. This should be compared with the 4246 channels available on the TAT7 coaxial cable link. Control circuitry enables a spare cable to be switched in if one of the active cables fails. As well as having spare cables, provision is made to switch in stand-by lasers (a photodiode placed on the non-emitting facet provides a measure of laser health). To increase system reliability further, redundant circuits are included in each regenerator.

This technology has been displaced on long-haul routes by the invention of fibre amplifiers, dense wavelength division multiplexing (DWDM) and low-loss dry fibre. Such a scheme is the Asia-Africa-Europe-1 (AAE-1) cable. The link is 25,000 km long, carries 100 Gbit/s data per channel (40 off) and uses five fibre pairs to give a total bit rate of 40 Tbit/s. The use of fibre amplifiers means that full regeneration is not required. Instead the attenuation of a length of fibre is compensated for by the amplifier. Although the link is expensive, it should be remembered that the capacity means the cost per unit of data will be small.

The system configuration of TAT8 is also used in short-haul industrial links, an example of which we will look at next. The system is basically an optical LAN, with a host computer controlling remote equipment. The maximum distance between terminals is typically 600 m, which is determined by physical constraints. This maximum distance, together with the 100 Mbit/s data rate, means that 62.5 μm core, 100 MHz km, PCS fibre can be specified. The attenuation at the 820 nm operating wavelength is typically 8 dB/km.

Packaged VCSELs with a typical output power of −3 dBm are used as the sources. At the receiver, a packaged Si PIN photodiode supplies a signal current to a transimpedance preamplifier. The receiver sensitivity is approximately −27 dBm, which results in an allowable link loss of 24 dB. As the fibre attenuation is 8 dB/km, and each link uses two connectors with a typical loss of 3 dB, the operating margin over a 600 m length is approximately 11 dB. The excess of received optical power means that a pre-detection filter is not required.

As the communications system is a LAN, data may have to be sent through a large number of terminals before it reaches the destination terminal. In view of this, each terminal extracts a clock from the data and regenerates the signal. The link controllers are housed in readily accessible locations, and so maintenance of the equipment is not a major problem. However, if the error rate over a particular link increases owing to increased fibre attenuation, reduced VCSEL power or lower receiver sensitivity, the entire network could collapse. To maintain transmission, a back-up link is installed. By comparing the synchronisation code in the received data frame to that of the ideal code, the error rate over the main link can be monitored. If errors are present, then data transmission can be automatically switched to the back-up link.

7.3 Long-Haul High-Data-Rate Links

It is fair to say that a revolution has occurred in optical fibre communications: regenerators are no longer needed every 30–40 km; dispersion is reduced; bit rates are in excess of 40 Gbit/s; dense wavelength division multiplexing (DWDM); and techniques that were only used in radio are now commonplace. In this section we will look at these advances and discuss their impact.

7.3.1 Optical Fibre Transmission Bands

As we have seen in Chap. 2, there are three transmission windows at 850, 1300 and 1550 nm. Between these regions of low loss are attenuation peaks due to the presence of water in the fibre. It is now possible to manufacture fibre in which the water concentration is less than 1 ppb – so-called dry fibre. With such a low concentration, the absorption peaks are vastly reduced, and it becomes possible to allocate bands of wavelength as shown in Table 7.2.

Table 7.2 Allocation of
bands in dry fibre

Window	
O – Original	1260–1360 nm
E – Extended band	1360–1460 nm
S – Short wavelength	1460–1530 nm
C – Conventional	1530–1565 nm
L – Long	1565–1625 nm
U – Ultra long	1625–1675 nm

Fig. 7.1 Four-level QAM
signal constellation

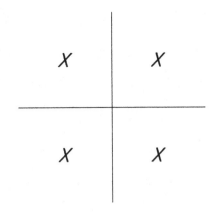

Within these bands are individual, 40 Gbit/s, channels based on a 50 GHz spacing. This results in 1180 channels giving an overall data rate of 47.2 Tbit/s down one fibre. This is referred to as dense wavelength division multiplexing (DWDM).

It may be wondered why we need such capacity. The answer lies with our use of information. We are now using data for Voice over Internet Protocol (VoIP), video calls and performing searches. All of this requires a fast communications link. Indeed, the data centre may be located in another continent.

7.3.2 Advanced Modulation Techniques

The 50 GHz channel spacing places a limit on the bit rate that can be transmitted. A similar thing happened years ago when the radio spectrum became crowded. The solution then was to use multi-level modulation such as quadrature amplitude modulation (QAM), and the same modulation scheme can be used with optical fibre.

Figure 7.1 shows the principle in which four QAM symbols are coded from two bits of data, so reducing the speed of transmission and hence bandwidth. Conversely, the bandwidth can remain the same, but the data rate can be increased. Higher coding orders are possible with 64 QAM being used. Modulation is quite easy as $LiNbO_3$ modulators can create accurate phase shifts. Demodulation uses the coherent detection technique described in Sect. 7.3.4.

7.3.3 Fibre Amplifiers

The invention of fibre amplifiers revolutionised long-haul optical fibre links. Such links are generally attenuated limited with repeater spacings of 30–40 km. However, with the advent of in-line fibre amplifiers, it became possible to boost the signal and compensate for the fibre losses. The link will then be dispersion limited, and the use of dispersion-flattened fibre and lasers with very small linewidth will give very long link lengths. Fibre amplifiers are also useful when using DWDM. In order to regenerate DWDM using electronic means, each individual channel would have to be detected, amplified, retimed and regenerated. Such a scheme would be impractical.

Figure 7.2 shows the schematic of a rare-earth-doped fibre amplifier. This is similar to the fibre laser considered in Sect. 3.8. If the fibre is doped with erbium atoms, Er^{3+}, the amplifier is denoted an erbium-doped fibre amplifier, EDFA. A pump, an 870 nm laser, excites the Er^{3+} electrons to a high energy state from which they rapidly drop to the required level. A population inversion is generated, and stimulated emission occurs with light in the C band. Gains of 20–30 dB are typical.

As well as gain, electronic amplifiers introduce noise, and EDFAs are no different. Noise is characterised by parameters known as the noise factor, F, and Noise Figure, NF, which is the noise factor in dB. The noise factor is defined by

$$F = \frac{S/N_{\text{in}}}{S/N_{\text{out}}} \tag{7.6}$$

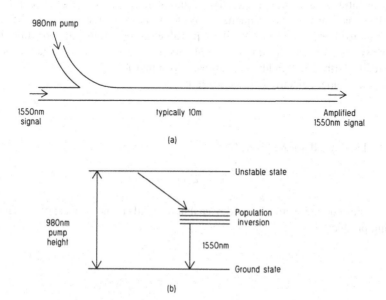

Fig. 7.2 (a) Schematic of an EDFA and (b) energy levels in the EDFA

and it is always greater than one. As the signal passes through the EDFA, the signal and the input noise are both boosted by the gain. However, there is also extra noise added by the amplifier itself. In an optical amplifier, this noise is due to spontaneous emission which is amplified – so-called amplified spontaneous emission, ASE.

Rare-earth-doped fibres can be used in cascade to compensate for the fibre losses. As an example, consider an EDFA with a gain of 20 dB. This can be used to compensate for total losses (splices and fibre) of 20 dB. Another amplifier can be used when the signal is attenuated by a further 20 dB. The effects of ASE are cumulative as every amplifier introduces it. In common with radio systems, we can use Friis' formula to calculate the resultant S/N at the end of the link:

$$F_{total} = 1 + (F_1 - 1) + \frac{(F_2 - 1)}{G_1} + \text{etc} \tag{7.7}$$

The minimum theoretical noise factor of an Er^{3+}-doped fibre amplifier is 2 (3 dB), and so the signal-to-noise ratio, SNR, is degraded by 2 every time it is amplified. If the distance between the optical amplifiers is such that the gain of each amplifier exactly matches the fibre loss between them, the SNR falls as the inverse of the transmission distance.

7.3.4 Coherent Detection

Coherent detection is commonly used in radio systems; the incoming signal is mixed with the output of a local oscillator. The resultant mixing products consist of the sum and difference of the two frequencies. With the advent of very narrow linewidth lasers (considered in Chap. 3), it is possible to use coherent detection using lightwave frequencies. As this is spatial mixing, the state of polarization is very important. (Orthogonal fields will result in no signal.)

Let the incident electric field be given by

$$E_{in} = E_s \cos \omega_s t \tag{7.8}$$

and the local oscillator be given by

$$E_{lo} = E_1 \cos \omega_1 t \tag{7.9}$$

The photodiode is a non-linear device that responds to the optical power. Thus the mixing products are

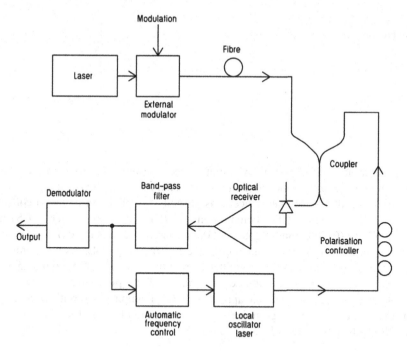

Fig. 7.3 Schematic of an optical heterodyne detection system

$$
\begin{aligned}
(E_{\text{in}} + E_{\text{lo}})^2 &= (E_s \cos \omega_s t + E_1 \cos \omega_1 t)^2 \\
&= E_s^2 \cos^2 \omega_s t + E_1^2 \cos^2 \omega_1 t + 2 E_s E_1 \cos \omega_s t \cos \omega_1 t
\end{aligned}
\tag{7.10}
$$

Expansion of the last term in (7.10) yields a very useful term

$$
E_s E_1 \cos (\omega_s - \omega_1) t
\tag{7.11}
$$

The other mixing terms are either d.c. or at a very high frequency. (In radio terms, the difference frequency in (7.11) is called the intermediate frequency, I.F.) It is worth noting that the signal amplitude has been amplified by the local oscillator, and so the sensitivity has been increased (Fig. 7.3).

7.3.5 Wideband Preamplifiers

We considered preamplifier design in Chap. 6. There we considered two basic designs – open-loop and feedback amplifiers. Feedback amplifiers can be used at high data rates, but there is a problem with stability due to two poles within the feedback loop.

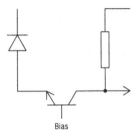

Fig. 7.4 Schematic of a CB input preamplifier

An alternative is to use a low input impedance amplifier such that the time constant associated with the front-end is very small giving a high bandwidth. As a transistor is a three-terminal device, it is possible to operate in common-collector (CC) (common-drain – CD), common-emitter (CE) (common-source – CS) and common-base (CB) (common-gate – CG). Of these three, it is the CB (CG) configuration that has the lowest input impedance of $1/g_m$ (Fig. 7.4). As an example, a bipolar transistor biased at 2 mA emitter current has a g_m of 0.08 S giving an input resistance of 12.5 Ω. A photodiode capacitance of 0.6 pF yields a bandwidth of 21 GHz. (A similar calculation applies to CG preamplifiers.) With such a low transimpedance, second-stage noise can become a problem, and so the following stages must be designed with care.

7.3.6   Optical Solitons

Optical solitons are produced by the interaction of high-energy optical pulses and certain non-linear effects in optical fibre. Their special characteristic is that they retain their shape for many hundreds if not thousands of kilometres. Thus an optical communications link that uses solitons will not suffer from dispersion. Obviously such a link would be very desirable, but before we examine soliton generation and propagation in optical fibre, let us examine the historical background of solitons in general.

In Victorian times, Scott Russell observed a 'solitary wave' travelling along the Union Canal in Scotland. Russell followed the wave for several miles, observing that the wave did not dissipate as normal waves should do. Unfortunately Victorian mathematics was not sufficiently advanced to explain the propagation of this wave. It was not until the late twentieth century that mathematics becomes advanced enough to solve the non-linear equations that governed Russell's wave. The solutions to these equations were termed *solitons*, and it was in 1972 that two Soviet theoretical physicists predicted the possibility of optical solitons (Zakharov and Shabat [2]).

Optical solitons in fibre are produced by the interaction between the material dispersion and a non-linear variation in refractive index. As we have already seen, the material dispersion passes through zero at a wavelength of 1.3 μm. Above this wavelength, shorter-wavelength signals tend to travel faster than longer-wavelength

Fig. 7.5 Soliton pulse in an optical fibre

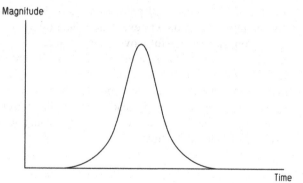

ones, and so the longer-wavelength components of an optical pulse appear in the trailing edge of the pulse and the pulse is dispersed. In a soliton, this broadening of the pulse is exactly balanced by the effects of a non-linear refractive index.

Let us consider a bell-shaped pulse, of the form shown in Fig. 7.5, propagating through a length of optical fibre. If the wavelength of operation is greater than 1.3 μm, the trailing edge of the pulse contains the slower long-wavelength pulse components. Taking the non-linear refractive index to be given by

$$n = n_o + n_2 I \tag{7.12}$$

where I is the intensity of the pulse, we find that a wave of constant amplitude undergoes a phase shift per unit length of

$$\delta\varnothing = \frac{2\pi}{\lambda_o} n_2 I \tag{7.13}$$

We can differentiate (7.13) with respect to time to give

$$\frac{d\delta\varnothing}{dt} = \frac{2\pi}{\lambda_o} n_2 \frac{dI}{dt} \tag{7.14}$$

Now, on the leading edge of the pulse, dI/dt is positive which results in the phase lag increasing as time goes by. Thus the leading edge is effectively slowed down. However, on the trailing edge, dI/dt is negative, and this results in a decreasing phase lag. Thus the trailing edge is speeded up. We can therefore see that the pulse is effectively shrunk by the non-linearity in refractive index (this is known as the *Kerr effect*). If this pulse shrinkage balances out the pulse dispersion, then we have a soliton pulse which is able to propagate for very large distances.

In order to send data using soliton pulses, the pulse shape should be $sech^2$ in form (shown in Fig. 7.5). However, any reasonably shaped pulse that has an area, R, that satisfies the inequality $R_o/2 < R < 3R_o/2$ will eventually evolve into a soliton with area R_o, and so specialised pulse shaping is not required. Any optical power not propagating as a soliton will eventually die away as a result of dispersion effects.

To propagate successfully over long distances, the amplitude of the soliton must be maintained so that the Kerr effect is continually present. However, optical signals suffer from attenuation, and so the solitons have to be regenerated after a certain distance. As dispersion is not a problem with soliton links, there is no need to regenerate the pulses fully. Instead, fibre amplifiers (refer to Sect. 3.8) can be used. In one experiment, WDM was used to carry a data rate of 1 Tbit/s over a distance of several thousand kilometres.

7.4   Free-Space Optical Communications

Most optical links use optical fibre; however, there is an alternative in the form of free-space optical communications. Transceiver units use several VCSEL transmitters operating at 850 nm or 1.55 µm and a sensitive receiver using the appropriate detector. It is usual to have a large collecting lens and a large area photodiode in order to collect the maximum amount of power.

Some advantages of free-space optical communications are licence-free operation and ease/cost of installation – it is simply a matter of placing units facing each other. Usual applications are to provide building-to-building links – WLAN to WLAN. Some schemes operate in a hybrid system that uses a microwave radio link as back-up should the optical link fail due to adverse weather conditions (mainly fog). Spatial diversity is needed because of scintillations (changes in refractive index due to temperature gradients that cause the received signal to move around the detector). The maximum reported data rate under clear sky conditions is 10 Gbit/s with a range of 2.5 km. However, an r.f. back-up link reduces the bit-rate to just 10 Mbit/s under extreme weather conditions.

7.5   Future Trends

In this section, we will consider some of the latest advances in optical communications. Most of our study will be descriptive in form, and we begin by examining optical fibres which exhibit very low loss at wavelengths above 1.55 µm.

7.5.1   Fluoride-Based Optical Fibres

One of the latest advances in optical fibres is the development of single-mode optical fibres, which exhibit very low loss in the mid infrared region, above 2 µm. As we saw earlier, Rayleigh scattering reduces as wavelength to the fourth power, and so very low-loss transmission requires operation at long wavelengths. However, in silica fibres, the absorption increases rapidly for wavelengths above the 1.55 μm window, and so very low-loss fibres have to be made from different materials.

The most promising glasses for low-loss fibres are those based on fluoride compounds. Of these, zirconium fluoride, ZrF_4, and beryllium fluoride, BeF_2, glasses have projected attenuations at 2.5 μm of 0.02 dB/km and 0.005 dB/km, respectively. Unfortunately, BeF_2 is highly toxic, and so most of the work has been concerned with ZrF_4 glasses. Probably the most suitable composition for fibre drawing is ZrF_4–BaF_2–LaF_3–AlF_3–NaF, usually abbreviated to ZBLAN. In ZBLAN fibres, the 2.87 μm fundamental of the OH bond causes a high level of attenuation. However, there is a transmission window at just over 2 μm, in which a measured loss of 0.01 dB/km has been recorded. Investigations show that the major loss mechanism is scattering from imperfections formed in the fibre during manufacture. So, with a more refined process, attenuations close to the Rayleigh scattering limit should be achievable.

The dispersion of ZBLAN fibres is highly dependent on the fibre structure. With an index difference of 0.014 and a core diameter of 6 μm, the dispersion is about 1 ps/nm/km, whereas a fibre with an index difference of 0.008 and a core diameter of 12 μm has a dispersion of greater than 15 ps/nm/km. By themselves, these dispersion times are very low; however, these fibres are likely to be used to transmit signals over very long distances, and so the total dispersion could be significant.

The move to higher wavelengths is likely to see a new generation of lasers and detectors. The most promising semiconductor laser source is a double heterojunction SLD, based on lnAsSbP matched to an InAs substrate. InGaAs photodiodes can operate at long wavelengths, but lead sulphide, PbS, detectors also show promise.

7.5.2 Graphene Detectors

Graphene is a new material with some rather special properties one of which is extremely useful for optical communications – a photodetector made from the material can respond to light of wavelength up to 6 μm. Unfortunately, the absorption coefficient of graphene is very low when illuminated by light normally incident to the surface of the material. (This is because graphene is a two-dimensional structure, and so only approximately 2% of the incident light is absorbed.) Alternatives are to confine the light in a Fabry-Perot resonator so that the light passes under the graphene many times, or to use a waveguide structure so that the light is gradually absorbed.

The structure of a graphene photodiode is similar to the MSM detector of Sect. 4.4. An interdigitated structure minimises the terminal capacitance to give a bandwidth in excess of 50 GHz. As mentioned previously, front illumination will not give a particularly good response due to the two-dimensional nature of the material. Thus, alternative structures, such as a silicon waveguide, are being considered as is the possible use in integrated optics.

7.5.3 Optical Wireless

The most common form of optical wireless being researched is known as visible light communication (VLC). This uses white light LEDs and pulses them at a high speed. The speed is high enough such that the brightness of the LED is not affected.

One of the problems with pulsing a white light LED is that the phosphor used to generate the white light introduces a time constant that limits the maximum modulation speed. This can be countered to some extent by introducing a speed-up transient as used in the circuit of Fig. 3.26. Speeds are variable, but 100 Mbit/s is achievable.

7.5.4 Crystalline Fibres

Photonic crystal fibre (PCF) is a type of fibre that has a solid glass core surrounded by hollow tubes containing air (Fig. 7.6). Propagation is in the high-refractive index core with the air-filled hollow tubes providing the cladding (ref). The technology is still in its infancy, but the applications are many: fibre lasers; fibre amplifiers; dispersion compensators; single-mode operation over a variety of wavelengths; and polarisation maintaining fibres. At present the fibres do not have the low loss needed for telecommunications systems.

7.5.5 Spatial Division Multiplexing (SDM)

Signal transmission in optical fibre uses time division multiplexing, wavelength division multiplexing, multi-phase division multiplexing and multi-level signalling. In fact, every electrical parameter that can be modulated has been used. One multiplexing system that is currently being explored is spatial division multiplexing (SDM) (see Fig. 7.7). In this technique, several cores are used in one fibre (multi-core fibre – MCF) with the cladding being the same as SM fibre. The fibres can be sufficiently separated so that cross-talk is kept to a minimum (distance between cores of >30 μm), although it is possible to work with closely spaced cores (distance

Fig. 7.6 Cross-section diagram of PCF

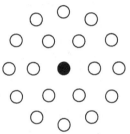

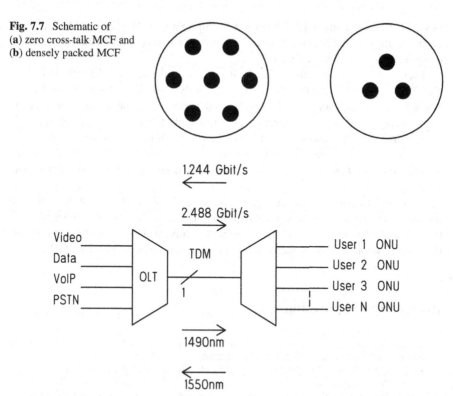

Fig. 7.7 Schematic of
(**a**) zero cross-talk MCF and
(**b**) densely packed MCF

Fig. 7.8 Schematic of a PON

between cores of <30 μm) such that cross-talk is intentionally introduced. In this case techniques used in Multiple In Multiple Out (MIMO) are used to separate the individual channels.

The maximum number of cores for a standard 125 μm cladding is 30. This together with a WDM system gives a fibre capacity in excess of 1 Pbit/s. Clearly there is a lot to be gained by SDM.

7.5.6 Passive Optical Networks (PONs)

One way of delivering fibre to the home is to have a dedicated line coming from a central terminal for each individual home (say 32). That would mean 32 fibre links and 64 transceivers (one at each end of the fibre). This would be expensive. An alternative is to use a Passive Optical Network (PON) to deliver services to the consumer through Fibre To The Home (FTTH). A PON (Fig. 7.8) consists of an Optical Line Terminal (OLT) that combines the services and produces a single data stream for modulating a 1490 nm laser. The combined output is then split using a $1 \times N$ splitter with each arm going to the N consumers (Optical Network Unit –

ONU). Up-link data uses Time Division Multiplexing (TDM) and a 1550 nm laser. In this manner, N users can be connected to a single fibre – far more efficient than using a single fibre for each user.

There are several difficulties with such a scheme – the distance from the splitter to the ONUs can be different, and this requires a dynamic range of typically 20 dB. Secondly, the data from the ONU is allocated a particular time slot as time division multiplexing (TDM) is used. Thus user 1 transmits for a time T second and then waits, while the other users access the line before it can transmit again. In the worst-case scenario, only user 1 is active and so data appears for a fraction of the time (T second in $32T$ second). All the electronics used must respond to this data sequence.

There is considerable interest in providing optical links for the last mile – the final link to the consumer premises. User requirements are for higher and higher bit rates to carry more bandwidth intensive services. The advent of VLC-based communications will, inevitably, mean increased demand which could be met by PONs.

Recommended Reading

1. Tamura Y et al (2018) The first 0.14-dB/km loss optical fiber and its impact on submarine transmission. IEEE Journal of Lightwave Technology 36(1):44–49
2. Zakharov VE, Shabat AV (1972) Exact theory of 2-dimensional self focusing and 1-dimensional self modulation of non-linear waves in nonlinear media. Soviet Physics - JETP 34:62–69
3. Infinera (2019) Infinera sets highest-performance 600G transmission record. https://www.infinera.com/wp-content/uploads/infinera-sets-highest-performance-600g-transmission_record.pdf. Accessed 4 Sept 2019
4. Mears RJ et al (1987) Low-noise erbium-doped fibre amplifier operating at 1.54µm. Electronics Letters 23(19):1026–1028
5. Hodgkinson TG et al (1985) Coherent optical transmission systems. British Telecom Technology Journal 3:5–18
6. Walker GR, Walker NG (1988) A rugged all-fibre endless polarisation controller. Electronics Letters 24:1353–1354
7. Brain M (1989) Coherent optical networks. British Telecom Technology Journal 1:50–57
8. Taylor JR (1992) Optical solitons: theory and experiment. Cambridge University Press, Cambridge, UK
9. Agrawal GP (1995) Nonlinear fiber optics, 2nd edn. Academic Press, San Diego, CA
10. France PW et al (1987) Progress in fluoride fibres for optical tele-communications. British Telecom Technology Journal 5:28–44
11. www.graphene-info.com
12. Ping M et al (2019) Plasmonically enhanced graphene photodetector featuring 100 Gbit/s data reception, high responsivity, and compact size. ACS Photonics 61:154–161
13. Knight JC (2003) Photonic crystal fibres. Nature 424:847–851
14. P. St.J. Russell (2006) Photonic crystal fibers. IEEE Journal of Lightwave Technology 24 (12):4729–4749
15. Saitoh K, Matsuo S (2016) Multicore fiber technology. IEEE Journal of Lightwave Technology 34:55–66
16. Razavi B (2012) Design of integrated circuits for optical communications, 2nd edn. Wiley, Hoboken, NJ

Index

A

Absorption coefficient, 155–156, 160
Acousto-optic modulator, *see* External
    modulators
Active region, 106
Analogue receivers, 204–205
    comparison of APD and PIN receivers,
      204–205
    noise equivalent bandwidth, 175, 203
    optimum avalanche gain, 205
    signal-to-noise ratio, 203
Anti-reflection coating, 157, 167
APD, *see* Avalanche photodiodes
Attenuation constant, 12
    See also Absorption coefficient
Avalanche breakdown, 154, 169
Avalanche photo diodes (APDs),
    154–174
    analogue receiver noise, 202
    breakdown, 154, 169
    current multiplication, 172
    dark current, 171, 175
    excess noise factor, 177
    ionisation coefficient, 172
    long-wave length devices, 171
    noise, 177, 197
    optimum gain, 199, 203
    primary current, 177
    reach-through, 169
    signal to noise ratio, 177
    speed, 173
    structures, 170
    zener breakdown, 169

B

Band-gap, 81, 86, 98–100
Bandwidth
    of optical fibre, 9, 65–67, 230–231
Bandwidth-length product, 10
Base-line wander, 201
Binding parameter, 38, 42
Bi-phase coding, *see* Manchester coding
Block coding, 201
Bragg cell, *see* Acousto-optic external
    modulators
Brewster angle, 136

C

Central decision detection, 192
Clock extraction, *see* Timing extraction
Coherent detection, 236
Common-emitter preamplifiers, 216–223
    See also Transimpedance preamplifiers
Common-collector preamplifiers, 223–227
    See also Transimpedance preamplifiers
Complementary error function, 194
Connectors, 74
Couplers, 75
Critical angle, 25
Cut-off frequency, 38, 58
Cylindrical coordinate set, 48

D

Dark current, 153, 157, 167, 175
DBR laser, *see* Semiconductor laser diode

© Springer Nature Switzerland AG 2020
M. Sibley, *Optical Communications*, https://doi.org/10.1007/978-3-030-34359-0

Dense wavelength division multiplexing
    (DWDM), 232
DFB laser, *see* Semiconductor laser diode
Digital receivers, *see* Optical receivers
Direct transitions, *see* Photons
Disparity, *see* Timing extraction
Dispersion
    material, 14–17, 43–45, 55, 61, 66
    modal, 17–21, 57–61, 66
    waveguide, 43–45, 54–57, 62–64, 67
Dispersion shifted fibre, 63

E

Eigenvalue equation, 31–32, 53,
    57, 58, 61
Electromagnetic spectrum, 2
Electro-optic effect, 143–150
Electro-optic modulator, *see* External
    modulators
ELED, *see* Edge emitting light emitting diodes
Equivalent input noise current spectral
    density, 185
Erbium-doped fibre amplifier, 235
Erfc, *see* Complementary error function
Error probability, 193–194
Evanescent wave, 27–28, 30, 34, 52, 64, 77
Excess loss, *see* Couplers
External modulators, 142–151
    acousto-optic bulk modulator, 148–151
    electro-optic bulk modulator, 145–151
    Mach-Zehnder interferometer, 143–148
    phase modulators, 143–148
Extinction ratio, 195, 205

F

Fabry-Perot etalon, 114, 138
Ferrule connectors, 76
Fibre amplifiers, 235
Fibre lasers, 151
Fluoride-based fibre, *see* Optical fibre
Free-space optical communications, 240
Fusion splicing, 75

G

Gaussian pulses, 65, 66, 197
Germanium photodiodes, 99, 155, 168,
    171, 178
Graded-index fibre, 9, 64, 72
Group refractive index, 19, 44, 45, 56
Group velocity, 19

H

HeNe lasers
    emission wavelengths, 137
Heterodyne detection, *see* Coherent detection
Heterojunctions, 106–108surface
    recombination, 113
    See also Semiconductor laser diode
High input impedance preamplifiers
    cascode input, 210
    compensation network, 210, 212
    design example, 215
    dynamic range, 215
    frequency response, 211–213
    noise analysis, 213–214
    voltage gain, 212
Hybrid modes, *see* Optical fibre

I

Indirect transitions, *see* Phonons
Industrial links, 3
Integrated optics
    electro-optic effect, 143–145
    lithium niobate, 77, 142
    Mach-Zehnder modulator, 144
    phase modulator, 143, 144
Integrating front-end receivers, *see* High input
    impedance preamplifiers
Inter-symbol interference, 165, 183
ISI, *see* Inter-symbol interference

J

Jitter, 188
    See also Timing extraction

L

Laser diode, *see* Semiconductor laser
Lasers
    fibre laser, 151
    gas, 138
    ruby, 1
    semiconductor, 2, 113–136
    solid-state, 135–139
    stimulated emission, 113, 115–125
Light emission in p-n diodes, 98–106
    band-gap, 98, 100
    heterojunctions, 106–108
    phonons, 99
    photons, 99
Light emitting diodes, 3, 108–113
    bandwidth, 112

drive circuits, 139, 140
edge emitting, 110
efficiency, 112
line-width, 111
spontaneous emission, 110
surface emitting, 109
Line-width, 20, 111
Lithium niobate, 77, 142–145
        See also Electro-optic effect; Integrated
            optics

M

Mach-Zehnder interferometer, see External
        modulators
Manchester coding, 200
Material dispersion, 19
        See also Dispersion, material
Maxwell's equations, 11, 32, 47
Meridional rays, see Optical fibre
Micro-bending, 65
Miller capacitance, 211, 219, 223
Modal dispersion, see Dispersion, modal
Mode-hopping, 124
Mode mixing, 65
Modes
        in a laser, 125, 126
        in an optical fibre, 47–54
        in a planar waveguide, 29–41
        See also Eigenvalue equation
Modified chemical vapour deposition, 71–72

N

Nd^{3+}:YAG lasers, 136, 137
        emission wavelengths, 137
Noise equivalent bandwidth, 175, 202
Non-radiative recombination, see Phonons
Non-return-to zero pulses, 190, 201
Non-zero extinction, see Extinction ratio
Normalised frequency, 32, 59, 77
Numerical aperture, 45, 46, 58

O

Operating margin, 230
        See also Power budget
Optical fibre, 2, 4, 9
        attenuation, 2, 75
        bandwidth, 9, 65–67
        bandwidth-length product, 10
        binding parameter 52
        connectors, 75

couplers, 75
cut-off frequency, 58
dispersion, 54–57, 61–64, 231
dispersion shifted, 63
double crucible, 74
eigenvalue equation, 53, 57–58, 61
electrical bandwidth, 65
electron absorption, 68
evanescent wave, 52, 64, 77
fibre pulling tower, 72
fluoride-based fibre, 240, 241
fusion splicing, 75
graded-index fibre, 9, 64, 72
hybrid modes, 54
impurity absorption, 68
material absorption, 70
material dispersion, 55, 61, 66, 238
materials, 71
Maxwell's equations, 48
meridional rays, 47
modal dispersion, 55–57, 60, 64–66
mode mixing, 65
modes, 48–54
modified chemical vapour deposition, 71
multimode fibre, 9, 57–61
normalised frequency, 58, 77
number of modes, 58–60
numerical aperture, 58, 61
optical bandwidth, 65–66
outside vapour phase oxidation, 71
plastic, 9, 70–71
plastic clad silica, 9, 70–71
power distribution, 60
preform fabrication, 71
propagation in, 46–64
rare earth doped, 151
Rayleigh scattering, 69
refractive index profile, 9
single-mode fibre, 9, 61–64
skew rays, 47
step-index, 9, 46–64
waveguide dispersion, 56–64, 67
weakly guiding fibres, 57
Optical fibre transmission bands, 233
Optical receivers, 183–206
        analogue, 202
        bandwidth, 212, 217, 225
        bandwidth integrals, 189–192
        base-line wander, 201
        block coding, 201
        central decision detection, 192
        comparison of PIN and APD detectors, 204
        error probability, 193

Optical receivers (*cont.*)
  extinction ratio, 195
  Illustrative of a non-zero extinction, 195
  inter-symbol interference, 183
  jitter, 188
  noise, 185–187
  noise equivalent bandwidth, 175
  non-zero extinction, 195
  photodiode noise, 197
  pre-detection filter, 185, 187
  sensitivity, 196, 205
  threshold crossing detector, 191
  timing extraction, 200, 232
  transimpedance, 184
  See also Preamplifiers
Optimum bias current, 220
Outside vapour phase oxidation, 71

P
Phase constant, 12
Phase velocity, 13, 14
  See also Propagation of light
Phonons, 98, 102
Photoconductivity, 153, 155
Photodiodes, 5, 153–181
  APD, 5, 169–174, 177–180
  cut-off wavelength, 159
  dark current, 153, 157, 167, 175
  germanium, 99, 155, 168, 171, 178
  noise, 174–181, 197–199, 203–205
  penetration depth, 159, 162
  photovoltaic mode, 154
  PIN, 5, 162–169
  quantum efficiency, 158–160, 162,
     163, 167
  quantum limit, 175, 180
  responsivity, 154, 158, 172, 197
  signal current, 157
Photonic crystal fibre (PCF), 242
Photons, 99
  See also Light emission in p-n diodes
PIN photodiodes, 5, 162–169
  circuit model, 167
  depletion layer, 164
  junction capacitance, 165–167
  long-wavelength PIN photodiodes,
     168–169
  punch-through voltage, 164
  quantam limit, 175, 180
  rear-entry, 169
  signal-to-noise ratio, 174, 203

speed of operation, 165–167
structure, 163
PINFET, *see* High input impedance
  preamplifiers; Preamplifiers
Planar waveguide, 22–46
  binding parameter, 38, 42
  cut-off condition, 38
  eigenvalue equation, 31, 32
  evanescent wave, 27–29
  group refractive index, 44
  maximum number of modes, 32
  Maxwell's equations, 32
  modal analyisis, 32–40
  modal dispersion, 40–42
  normalised frequency, 32
  numerical aperture, 45
  propagation Modes, 29
  single-mode, 39–40
  standing waves, 34
  waveguide Dispersion, 43–45
Plastic clad silica, *see* Optical fibre
Plastic fibre, *see* Optical fibre
Population inversion Ill, 115, 119,
     129, 136, 151
Power budget, 229
Power flow, *see* Poynting vector
Poynting vector, 15
Preamplifier design, 237
Preamplifiers
  PINFET, 3, 210–215
  transimpedance, 5, 217–224
Pre-detection filter, 185, 187
Propagation coefficient, 12
Propagation modes, 29
  See also Modes
Propagation of light
  group refractive index, 19,
     44, 56
  group velocity, 17, 43
  impedance, 15
  in a dielectric, 10
  Maxwell equations, 32, 48
  Maxwell's equations, 11
  phase velocity, 14–18
  planar waveguide, 22–46
  power flow, 15, 27
  Poynting vector, 15
  propagation parameters, 13
  reflection and refraction, 22–29
  reflection coefficient, 23
  Snell's Laws, 24
  transmission coefficient, 23
  wave equation, 11

Q

Quantum efficiency, 112–113, 158–160, 162,
    163, 167
Quantum limit, 175–177, 180
Quantum noise, 175–177

R

Radiative recombination, *see* Semiconductor
    diode, recombination; Photons
Raised-cosine spectrum pulses, 187
    See also Pre-detection filter
Rare earth doped fibre, 151, 236
Rate equations, *see* Semiconductor laser diode
Receiver noise
    analogue, 202–203
    comparison of APD and PIN detectors, 204,
        205
    digital, 186–203
    input noise current, 185–187
    measurement of sensitivity, 205
    prediction of sensitivity, 206, 207
Reflection coefficient, 23
Refractive index, 14
    effective, 38
Responsivity, 154, 158, 172, 197

S

Semiconductor diode, 79–98, 153–181
    avalanche breakdown, 154, 169
    band-gap, 81, 86, 98
    barrier potential, 88, 91, 160
    carrier concentration, 91
    carrier density, 92, 94, 97
    carrier lifetime, 95, 118
    current density, 94
    depletion region, 86, 91, 160
    diffusion length, 96
    direct transitions, 98, 100
    electron–hole product, 81
    Fermi level, 80
    forward bias, 93
    heterojunction, 106
    indirect transitions, 98, 100
    intrinsic material, 79
    light emission, 98
    recombination, 98, 102
    zener breakdown, 169
    zero bias, 86
Semiconductor laser diode, 113–135
    bandwidth, 151
    buried heterostructure, 133

DBR, 134
DFB, 134
drive circuits, 140
gain profile, 128
heterojunction, 106–108
linewidth, 138
mode-hopping, 124
modes, 126
packaging, 109
population inversion, 115
rate equations, 100
reliability, 232
spontaneous emission, 116
stimulated emission, 113, 115–125
Semiconductor material
    diffusion current, 84
    diffusion length, 85
    drift current, 84
    drift velocity, 84
    electron mobility, 84
    extrinsic material, 83
Sensitivity, 196, 205–206
Signal-to-noise ratio, 174, 203
Single-mode optical fibre, 61–64
Single-mode planar waveguide, 39
Skew rays, 47
SLD, *see* Semiconductor laser diode
SLED, *see* Surface emitting Light emitting
    diodes
Snell's law, 24, *see* Propagation of light
Solitons, 238
Spontaneous emission, 110, 116
Step-index fibre, 47–64
Stimulated emission, 113, 115–125
System design, 229–231
    link capacity, 231
    power budget table, 229
    timing extraction, 232

T

Threshold crossing detector, 191
Timing extraction, 200–202
Total internal reflection, 25
Transimpedance, 184
Transimpedance preamplifiers, 209, 216–223
    circuit, 217
    design example, 222, 225
    dynamic range, 214
    frequency response, 218, 224
    noise analysis, 220, 225
    voltage gain, 217, 222
Transmission coefficient, 23

W
Wave equation, *see* Propagation of light
Waveguide dispersion, 56–64, 66
    See also Dispersion,waveguide
Wavelength division multiplexing (WDM), 240
Weakly guiding waveguides, 42, 44, 55

Z
Zener breakdown, 169

Printed in the United States
By Bookmasters